Henry Law, D. Kinnear Clark

The construction of roads and streets

In two parts

Henry Law, D. Kinnear Clark

The construction of roads and streets
In two parts

ISBN/EAN: 9783742877772

Manufactured in Europe, USA, Canada, Australia, Japa

Cover: Foto ©berggeist007 / pixelio.de

Manufactured and distributed by brebook publishing software (www.brebook.com)

Henry Law, D. Kinnear Clark

The construction of roads and streets

THE CONSTRUCTION
OF
ROADS AND STREETS.

THE CONSTRUCTION
OF
ROADS AND STREETS

IN TWO PARTS

I.—THE ART OF CONSTRUCTING COMMON ROADS
By HENRY LAW, C.E.
Revised and Condensed by D. Kinnear Clark, C.E.

II.—RECENT PRACTICE IN THE CONSTRUCTION
OF ROADS AND STREETS
INCLUDING PAVEMENTS OF
STONE, WOOD, AND ASPHALTE

By D. KINNEAR CLARK, C.E.
MEMBER OF THE INSTITUTION OF CIVIL ENGINEERS; AUTHOR OF "RAILWAY
MACHINERY," "A MANUAL OF RULES, TABLES, AND DATA," ETC.

With numerous Illustrations

LONDON
CROSBY LOCKWOOD & CO.
7, STATIONERS' HALL COURT, LUDGATE HILL
1877

LONDON:
PRINTED BY VIRTUE AND CO., LIMITED,
CITY ROAD.

PREFACE.

THE present work consists of two parts. The first comprises "The Art of Constructing Common Roads," by Mr. Henry Law, revised and condensed; the second consists of "The Recent Practice in the Construction of Roads and Streets," by Mr. D. Kinnear Clark, C.E., in the investigation of which he has been indebted for much material to the excellent Reports of Lieutenant-Colonel Haywood, Engineer and Surveyor to the Commissioners of Sewers of the City of London. The whole is preceded by an historical sketch of the subject, also by Mr. Clark.

The City of London is a microcosm of the best and most varied experience in carriage-way construction, under the superintendence of the Engineer who has lucidly described the various structures which have, from time to time, been laid down and tried, in a catholic spirit, and has recorded the results of his experience, in a series of Reports ranging over a period of thirty years —from 1848 to 1877. Mr. Clark has endeavoured impartially to set forth the merits and disadvantages of the systems of pavement which have come under his observation, and he believes that the results of his investigations will be useful to others.

The varieties of wood pavement and of asphalte pavement which have been laid in the Metropolis— more especially in the City—have been fully described and, it is hoped, fairly criticised. Mr. Clark has also added a chapter on the Resistance to Traction on Common Roads, in which he has endeavoured to educe the law of rolling resistance, and has contributed new formulas, with fresh data.

Appended to the text will be found a portion of a paper by Sir John Burgoyne on Rolling New-made Roads; some valuable extracts from Mr. Frederick A. Paget's Report on Road-rolling, containing several interesting historical facts; and finally, a table showing the Condition of Wood and Asphalte Carriage-way Pavements in the City of London, from a recent Report of Colonel Haywood.

CONTENTS.

HISTORICAL SKETCH. By D. K. CLARK.

Country Roads.—Barrelled Roads.—Macadam's Roads.—Telford's Roads.—Length of Metalled Roads, in 1868-69.—Boulder Pavement.—London Pavements.—Wood Pavements in Russia.—Wood Pavements in the United States.—Stead's Wood Pavement.—De Lisle's Wood Pavement.—Carey's Wood Pavement.—French Roads 1

PART I.—CONSTRUCTION OF ROADS.
By HENRY LAW, C.E.

CHAPTER I.—EXPLORATION FOR ROADS:—Principle of Selection of Route.—Contour Lines.—Taking Levels.—Bench-marks.—Sections.—Laying out a Road 21

CHAPTER II.—CONSTRUCTION OF ROADS: EARTHWORK AND DRAINAGE:—Earthwork.—Trial Pits.—Working Plan.—Cuttings and Fillings.—Side Slopes.—Excavation in Rock.—Slips.—Drainage.—Embankments.—Catch-water Drains.—Road on the Side of a Hill.—Side-cuttings.—Spoil-bank . 40

CHAPTER III.—RESISTANCE TO TRACTION ON COMMON ROADS:—M. Morin's Experiments.—Sir John Macneil's Experiments.—Resistance on Inclines.—Table of Resistance on Inclined Roads.—Professor Mahan's Deductions.—Angle of Repose . 51
NOTE BY THE EDITOR:—Sir John Macneil on Gradients.—Professor Mahan on Gradients.—M. Dumas on Gradients.—M. Dupuit on Gradients 63

CHAPTER IV.—ON THE SECTION OF ROADS:—Gradients.—Table of Gradients and Angles of Roads.—Width and Transverse Section of Roads.—Mr. Macadam's Views.—Mr. Walker's Views.—Proposed form of Cross Section.—Professor Mahan's Views.—Form of the Bed.—Mr. Hughes's Views.—Drainage of Roads.—Formation of Drains.—Footpath.—Drainage for Marshy Soils 65

CHAPTER V.—ON REPAIRING AND IMPROVING ROADS:—Improvement of the Surface.—Best Season for Repairs.—Formation of Mud.—Watering Roads.—Tools Used.—Scraping Machines 79

CHAPTER VI.—CONSTRUCTION OF ROADS: FOUNDATION AND SUPERSTRUCTURE:—Soft Foundations.—Classification of Roads.—Solidity.—Foundations of Concrete.—Mr. Penfold's Practice.—Binding.—Mr. Telford's Practice in Foundations.—Covering.—Cementing or Solidifying the Surface.—Angular Stones.—Mr. Macadam's Practice.—Mr. Telford's Practice.—Gravel.— Mr. Hughes's Practice. — Chalk Binding. — Faggots.—Mr. Walker on the Use of Iron Scraps for Binding 90

CHAPTER VII.—PAVED ROADS AND STREETS:—Excavation.—Stone Sets.—Curb.—Paving for Inclined Streets.—Sidewalks and Crossings 104

CHAPTER VIII.—ON HEDGES AND FENCES:—Different Kinds of Fences.—Dry Rubble.—Post and Rail.—Quickset Hedge.—Sir John Macneil and Mr. Walker on Close and High Hedging 110

CHAPTER IX.—ON TAKING OUT QUANTITIES FOR ESTIMATES:—Earthwork.—Table of Contents of Cuttings or of Embankments 114

PART II.—RECENT PRACTICE IN THE CONSTRUCTION OF ROADS AND STREETS.

BY D. K. CLARK, C.E.

CHAPTER I.—MATERIALS EMPLOYED IN THE CONSTRUCTION OF ROADS AND STREETS:— For Carriage-ways: Stones.—Granite.—Table of Crushing Resistance of Granite.—Table of Crushing Strength and Absorbent Power of Various Stones.—Trap Rocks.—Comparative Wear of Stones.—Table of the Relative Wear of Granites, &c.—For Footpaths: Table of the Composition, Specific Gravity, and Strength of Sandstones.—Mr. Newlands' Observations.—Asphalte.—Artificial Asphalte.—Table of the Crushing Resistance of Timber.—Mr. Hope's Experiments on the Wear of Wood . . . 122

CONTENTS. ix

PAGE

CHAPTER II.—CONSTRUCTION OF MODERN MACADAM ROADS:—
Boning Rods.—First-class Metropolitan Roads.—Second-
class Metropolitan Roads.—Country Roads 134

CHAPTER III.—MACADAMISED ROADS—WEAR:—Weak or "Elas-
tic" Roads.—Mr. John Farey on Wear of Roads.—Compa-
rative Action of Feet of Horses and Wheels of Vehicles.—
Sir John Macneil on the Four-horse Stage-coach, and on the
Weight of Vehicles and Width of Tyres on Common Roads.
—Mr. James Macadam on Weight of Vehicles and Width of
Tyres.—M. Dupuit on Width of Tyres.—Mr. Joseph Mit-
chell on the Proportion of Vacuity to Solid Material, in
Broken Stones.—Mr. Bokeberg on the same.—Mr. Mitchell's
Analysis of the Crust of a Macadam Road.—The Road-
roller.—Annual Wear of Metalled Roads 138

CHAPTER IV.—MACADAMISED ROADS—COST:—*Roads in London.*
Mr. F. A. Paget's Data, with Table.—Suburban Highways.
—Mr. George Pinchbeck's Data.—Local Roads . . . 151
Roads in Birmingham.—Mr. J. P. Smith's Data . . . 155
Streets and Roads in Derby.—Mr. E. B. Ellice-Clark's Data.—
Table of Macadamised Streets.—Table showing Estimated
Cost of Paving.—Table of Comparative Costs for Granite
and Macadam 158
Roads in Sunderland.—Mr. D. Balfour's Data 161
Roads in Districts near Edinburgh, Glasgow, and Carlisle.—Mr.
J. H. Cunningham's Data 162

CHAPTER V.—CONCRETE ROADS:—Mr. Joseph Mitchell's Concrete Macadam 163

CHAPTER VI.—MACADAMISED ROADS IN FRANCE:—M. Dumas'
Views.—Type Sections of Roads 165

CHAPTER VII.—STONE PAVEMENTS—CITY OF LONDON:—Con-
struction of Early Pavements.—Colonel Haywood's Reports.
—Table of Earliest Granite Pavements.—Mr. Kelsey on the
Cost for Reparation, with Table.—Introduction of Three-
inch Sets.—Colonel Haywood's Tables of the Lengths of
London Pavements in 1848, 1851, and 1866.—Mr. William
Taylor on the Euston Pavement.—Experimental Paving laid
by Colonel Haywood in Moorgate Street, with Tables.—
Granites that have been laid in the City of London.—Rota-
tion of Granite Paving.—Traffic in the City.—Duration of
Three-inch Granite Paving in the City, with Table.—Colonel
Haywood's Estimate of its Duration and Cost, with Table.
—Example of London Bridge.—Blackfriars Bridge.—Typi-

cal Sections and Plans of a Fifty-feet Street for the City.
—Southwark Street 169

CHAPTER VIII.—STONE PAVEMENTS OF LIVERPOOL:—Mr. Newlands on the Length of Pavement in 1851.—Tables of Cost for Construction of Set Pavements, Boulder Pavements, and Macadam 193

CHAPTER IX.—STONE PAVEMENTS OF MANCHESTER:—Early Boulder Pavements.—Mr. H. Royle on Set Pavements.—Use of Pitch Grouting.—Cost.—Macadam 198

CHAPTER X.—WEAR OF GRANITE PAVEMENTS:—In the City of London, with Tables.—Data for Wear and Duration, with Table 202

CHAPTER XI.—STONE TRAMWAYS IN STREETS:—Mr. Walker's Tramways in the Commercial Road.—Resistance on Stone Tramways.—Granite Tramways in Northern Italy.—Mr. P. Le Neve Foster, Jun.'s Data.—Prices of Work at Milan. 208

CHAPTER XII.—WOOD PAVEMENT:—Dimensions of Blocks.—Interspaces 215

CHAPTER XIII.—CAREY'S WOOD PAVEMENT:—Carey's Pavement in the City of London, with Tables of Cost and Duration.—Carey's most recent Practice 217

CHAPTER XIV.—IMPROVED WOOD PAVEMENT:—First Laid in the City of London.—Construction.—Asphalte Grouting.—Objections to the Flooring.—Most recent Practice . . 223

CHAPTER XV.—OTHER WOOD PAVEMENTS:—Ligno-mineral Pavement.—Asphaltic Wood or Copeland's Pavement.—Harrison's Wood Pavement.—Henson's Wood Pavement.—Norton's Wood Pavement.—Mowlem's Wood Pavement.—Stone's Wood Pavement.—Gabriel's Wood Pavement.—Wilson's Wood Pavement.—Table of Wood Pavements in the City of London 228

CHAPTER XVI.—COST AND WEAR OF WOOD PAVEMENTS:—Cost in the City of London.—Mr. G. J. Crosbie-Dawson's Data.—Mr. Ellice-Clark's Data 236
Wear in the City of London.—Relation of Wear to Traffic.—Table of Estimated Duration 238

CHAPTER XVII.—ASPHALTE PAVEMENTS.—First used in Paris.—Mode of Construction in the City of London.—Val de Travers Compressed Asphalte Pavement.—Val de Travers Mastic Asphalte Pavement.—Limmer Mastic Asphalte Pavement.—Barnett's Liquid Iron Asphalte Pavement.—

Trinidad Asphalte Pavement.—Patent British Asphalte Pavement.—Montrotier Compound Asphalte Pavement.—Société Française des Asphaltes.—Maestu Compound Asphalte.— Stone's Slipless Asphalte.— Bennett's Foothold Metallic Asphalte.—Lillie's Composite Pavement.—McDonnell's Adamantean Concrete Pavement. — Granite Pavements with Asphalte Joints.—Table showing the Extent of Asphalte Pavements in the City of London, 1873.—Colonel Haywood's Deductions from his Experience.—Table showing Duration and Repair of Asphalte Pavements in the City of London, at March, 1873.—Table showing the Wear of Asphalte Pavements in Proportion to Traffic.—Colonel Haywood's Conclusions as to the Durability of Asphalte Pavements.—Cost and Terms of Contracts for Asphalte Pavements, with Table.—Val de Travers Asphalte in Manchester 242

CHAPTER XVIII.—OTHER PAVEMENTS:—Metropolitan Compound Metallic Paving.—Cast-iron Paving.—Cellular-iron Pavement.—Artificial Granite Pavement.—Compound Wood and Stone Pavement.—Concrete Pavement 261

CHAPTER XIX.—COMPARISON OF CARRIAGE-WAY PAVEMENTS:— Comparative Costs of Pavements.—Cost of Yorkshire Paving-stones.—Foot Pavements.—Comparative Slipperiness, with Table.—Comparative Convenience.—Report of Committee of the Society of Arts 264

CHAPTER XX.—CLEANSING OF PAVEMENTS:—Composition of Mud.—Dr. Letheby's Analysis, with Table.—Moisture in Mud.—Cleansing by Machinery and by Manual Labour.— Proportion of Granitic Detritus in Dust from a Granite Pavement.—Comparative Cost of Cleansing Granite and Macadam.—Watering with Jet and Hose, with Table—Mr. J. Lovick's Experiments.—Cleansing in Paris . . . 268

CHAPTER XXI.—MOUNTAIN ROADS:—Principle of Selection of Route—Major James Browne's Data.—Mr. Dobson on Roadmaking in New Zealand.—Major Browne on the Cost of Roads in India, and Method of Construction . . . 282

CHAPTER XXII.—RESISTANCE TO TRACTION ON COMMON ROADS:— Investigation of Rolling Resistance on Impressible Roads.— Work in Compressing the Material.—Resistance is Inversely Proportional to the Cube Root of the Diameter.— M. Morin's Deductions.—M. Dupuit's Deductions.—M. Debauve's Data, with Table.—Resistance of M. Loubat's Omnibus.—Experiments by Messrs. Eastons and Anderson on

xii CONTENTS.

the Resistance of Agricultural Carts and Waggons.—Sir John Macneil's Experiments on the Resistance of a Stagecoach, with Table.—Formula for Resistance of a Stagecoach.—M. Charié-Marsaines on the Performance of Flemish Horses, withTable.—Mr. D. K. Clark's Data . . . 290

APPENDICES.

I. ON ROLLING NEW-MADE ROADS. By General Sir John F. Burgoyne, Bart. 301

II. EXTRACTS FROM "REPORT ON THE ECONOMY OF ROAD MAINTENANCE AND HORSE-DRAFT THROUGH STEAM ROAD-ROLLING, WITH SPECIAL REFERENCE TO THE METROPOLIS." By Frederick A. Paget, C.E. . . . 309

III. EXTRACT FROM THE REPORT OF COLONEL HAYWOOD, Engineer and Surveyor to the Commissioners of Sewers, City of London, ON THE CONDITION OF WOOD AND ASPHALTE CARRIAGE-WAY PAVEMENTS, ON THE 1ST FEBRUARY, 1877 324

INDEX 338

CONSTRUCTION
OF
ROADS AND STREETS.

HISTORICAL SKETCH.

BY D. K. CLARK, C.E.

In the middle of last century, communication between towns was difficult. The roads were originally mere foot-paths, or horse-tracks, across the country, and the few wheeled carriages in use were of a rude and inefficient description, for which the roads were wholly unadapted. The roads were necessarily tortuous, every obstacle which the ground presented being sufficient to turn the traveller out of his natural direction. Many of these roads were carried over hills to avoid marshes, which were subsequently drained off or dried up; others deviated from their direct course in order to communicate with the fords of rivers now passable by bridges. The inland commerce of the country was chiefly carried on by transport on the backs of pack-horses, and the old-fashioned term *load*, commonly in use as a measure of weight, is a remnant of that custom—meaning a horse-load. Gradually, the roads became practicable for the rude carriages of the times, and they were maintained, though in a very defective condition, by local taxes on the counties or parishes in which they were situated. So they remained until turn-pike-trusts were established by law, for levying tolls from persons travelling upon the roads. Several of these trusts were established previous to 1765, and they subsequently became general, when the attention of all classes of the community was directed to the state of the highways. Bills for making turnpike-roads were passed, every year, to an extent which seems almost incredible; and, in addition, every parish was compelled by the force of public opinion, supplemented by indictments and fines recoverable at common law against the trustees, when the roads were not maintained in proper repair. But the turnpikes formed a cumbrous system: they were trusts in short lengths— about fifteen or eighteen miles—and the surveyors employed appear to have been ill-educated, and were appointed by favour of the trustees rather than for any professional knowledge.

A long period elapsed before any good system of roadmaking was established. The old crooked horse-tracks were generally followed, with a few deviations to render them easy; the deep ruts were filled with stones or gravel of large and unequal sizes, or with any other materials which could be obtained nearest at hand. The materials were thrown upon the roads in irregular masses, and roughly spread to make them passable. The best of those roads would, in our time, be declared intolerable. Roadmaking, as a profession, was unknown, and scarcely dreamt of; for the people employed to make the roads and keep them in repair, were ignorant and incompetent for their duties. Travelling was uncommon, and funds were scanty, and higher talent could not be commanded. Engineers, except in cases of special difficulty, such as the construction of a bridge over a deep and rapid river, cutting through a hill, or embanking across a valley, probably thought that road-making was beneath their consideration, and it was thought singular that Smeaton should have condescended to make a road across the valley of the Trent, between Markham and Newark, in 1768. At the same time, civil engineers, according to Sir Henry Parnell, "had been too commonly deemed by turnpike-trustees as something rather to be avoided, than as useful and necessary to be called to their assistance." By-and-bye, as people became sensible of the value of time, easier and more rapid means of communication than the old roads were required: improved bridges were built with easier ascents; and, in some cases, cuts were made to shorten the distances, though the general lines of the old roads were preserved. The roads, no doubt, were somewhat improved in this way, but there was no general system or concert between the district trustees.

Mr. Arthur Young, in his "Six Months' Tour," published in 1770, writes of some of the roads in the north of England:—"*To Wigan.* Turnpike.—I know not, in the whole range of language, terms sufficiently expressive to describe this infernal road. Let me most seriously caution all travellers who may accidentally propose to travel this terrible country, to avoid it as they would the devil, for a thousand to one they break their necks or their limbs by overthrows or breakings down. They will here meet with ruts, which I actually measured four feet deep, and floating with mud only from a wet summer; what therefore must it be after a winter? The only mending it receives is tumbling some loose stones, which serve no other purpose than jolting a carriage in the most intolerable manner. These are not merely opinions, but facts; for I actually passed three carts, broken down, in those eighteen miles of execrable memory." "*To Newcastle.* Turnpike. —A more dreadful road cannot be imagined. I was obliged to hire two men at one place to support my chaise from overturning. Let me persuade all travellers to avoid this terrible country, which must either dislocate their bones with broken pavements, or bury them in muddy sand." Even so much later as the year 1809, the roads answered to the description of Mr. Young. Mr. C. W. Ward, writing in that year,* states that the convex section, as shown in Fig. 1, was the most prevalent in the

Fig. 1.—Common Convex Road, in 1809.

country. Under the impression that the higher the arch was made, the more easily the road would be drained, the materials were heaped up about the centre till the sides became dangerous, by their slope, for the passage of carriages. The carriages, therefore, ran entirely upon the middle till it was crushed and worn down, and then a fresh supply of materials was laid on, and the road was again restored to its dangerous shape. The sides of the road were but little used, except in summer, or until the heavy waggons had crushed the middle into a surface apparently compact and smooth. In some places, the rough materials were laid in a narrow line, not exceeding seven or eight feet in breadth, along the middle of the road, and the sludge collected from the scrapings of the roads or ditches was placed on each side, like banks, to prevent the stones from being scattered by the wheels. The high convex form was so exceedingly defective as to defeat the object for which it was constructed. Carriages were forced, for safety or for convenience, to keep to the middle, and it was speedily ploughed into deep ruts, which held the rain-water, even when the convexity approached to the form of a semicircle. The central elevation, therefore, was not kept dry; and the central pressure of the traffic forced the material upon the sides, where they lay loose

* Third Report from Parliamentary Committee on Turnpikes and Highways, 1809.

and unconnected, and obstructed the course of water from the middle. The condition of such a road, ploughed and disintegrated, is illustrated in section by Fig. 2, when it

Fig. 2.—An Indicted Road.—Its first state. Year 1809.

was, probably, indicted. It was common for the parish-surveyor after harvest to make a contract with a stout labourer, who took job-work, for the reparation of the road, with a special injunction "to be sure that he threw up the road high enough, and made the stones of the old causeway, or foot pavement, go as far as they could." The diligent operator fell to work; nor was he stopped by the equinoctial rains in September, for the work must be done, as contracted for, before the Michaelmas sessions. He accordingly produced something, Fig. 3. The clods

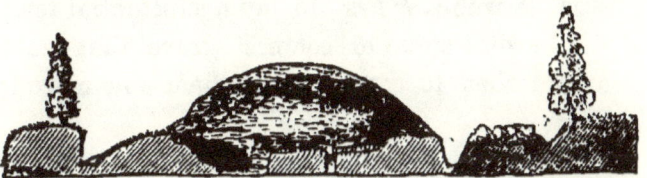

Fig. 3.—The Indicted Road thrown up, to take off the Indictment, under the direction of a Parish Surveyor. Its second state.

and rushes were thrown into the bottom, and the soft soil which nourished the vegetation, and all other materials, hard or soft, were laid down, forming a convexity of considerable elevation, according to order:—barrelling the road, as it was called. The whole was duly surmounted with the stones from the old broken footpath, with a little gravel raked over them, just to keep them together. Finished thus, say by Saturday night, then on the following Monday it was submitted for inspection to two magistrates, on their way to the quarter sessions. How could they possibly refuse to speak the truth? they certified " that it was perfectly smooth when they saw it, and that a vast deal had been done since the last time they were there." But besides tear and wear, decomposition immediately took place in the chaotic mass, and, in the second or third year after the repair, the road was reduced to the condition shown in Fig. 4, in its last and worst state.

Although it appears that the practice of road-making,

Fig. 4.—The same Road, in its third year after repair, or its last and worst state.

even at the commencement of the present century, was sadly deficient, it is, nevertheless, fair to add that persons of intelligence were aware of the first requisite for a good road. Mr. Foster, of Bedfordshire, in 1809, saw that it was desirable, " first, to lay a substantial foundation of the hardest stone or coarsest gravel that could be procured, and then to coat it with a finer and more level surface."

It followed, from the imperfect condition of the roads, that the wheels of vehicles were required to be of great width, in proportion to the weight carried on each wheel. The following table shows the proportions and the distribution of weight on the wheels, according to the regulations of the Act which was in force in the early part of this century. The rolling widths are the slant widths of conical wheels:—

TABLE No. 1.—WEIGHT, HORSE-POWER, AND WHEELS OF VEHICLES ON COMMON ROADS. 1809.

Breadth of wheel.	Gross weight.	Number of horses.	Draught of each horse.		Weight on the road at each wheel.	Pressure per inch of width.
inches.	tons.		cwt.	lb.	cwt.	lbs.
16 . . .	8	10	16	0	40	280
9, rolling 16 .	6½	8	16	42	32½	404
9 . . .	6	8	15	0	30	373
6, rolling 11 .	5½	6	18	37	27½	513
6 . . .	4½	6	16	0	22½	420
3 . . .	3¼	4	·17	56	17½	653
2, Stage coach .	4	4	20	0	20 ;	1120

Here, it is apparent that the pressures per inch of width of tyres increased as the width diminished. In the opinion of the practical men of that day—carriers and others— the pressure should have been limited to about 4 cwt., or 448 lbs., per inch wide; and it was maintained that the minimum width of wheel for any vehicle should be 4½ inches.

About the year 1816, Mr. James L. Macadam, who had for many years previously given his attention to the state of the roads, assumed the direction of the roads of Bristol, and he put in practice the leading principle of his system of road-making, namely, "to put broken stone upon a road, which shall unite by its own angles so as to form a solid, hard surface." "It follows," he adds, "that when that material is laid upon the road, it must remain in the situation in which it is placed without ever being moved again; and what I find fault with in putting quantities of gravel on the road is that, before it becomes useful, it must move its situation, and be in constant motion."[*]

The principle was to substitute small angular stones, such

[*] "Report of the Select Committee on the Highways of the Kingdom, 1819," p. 22.

as resulted from the breakage of larger stones, for rounded stones; so as to form a sort of mosaic or interlocking system. This is the distinctive novelty of the system of Macadam, and its value has been established by universal experience.

Mr. Macadam also maintained that no greater convexity should be given to the surface of the road, in transverse sections, than was sufficient to cause rain-water to run readily into the side channels. The surface of the road was kept even and clean by the addition of proper fresh materials when necessary, distributed equally in thin layers immediately after rain, in order that the new materials might bind and incorporate properly with the old. Macadam's system of construction consisted in simply laying a stratum of flints, or other hard materials, 10 or 11 inches thick, broken equally into small pieces about 2 inches in diameter, and spread equally over the intended road-surface. The broken "metal" became consolidated by carriages passing over it. Without any specialty of professional training, except the faculty of acute observation, Macadam effected great improvement of the surface of the roads immediately under his charge; and, by his business-like and extended views on road-administration, he established for himself a world-wide reputation. He professed to be a road-maker only, and he devoted his whole time and attention to the propagation of his system. He found the roads in the Bristol district loaded with two or three feet of materials, of large and irregular size, which had for years been accumulated on the surface. The heaps were utilised as quarries of stones partially broken on the spot; the stones he excavated, separated from the mud, and reduced by breakage to a uniform size, 6 ounces in weight. After having been so broken, the stones were relaid, and were carefully and regularly raked and levelled during the process of consolidation. In this way, with the addition of effective drainage where necessary, he was enabled to make a good surface on roads which previously were almost impassable. As nearly every road had more metal upon it than was necessary, he, and the surveyors appointed by him, established economy in the construction and maintenance, as well as in the administration of the finances, and his system became generally adopted.

Whilst Mr. Macadam deserved well as the pioneer of good road-construction, it may be observed that he had been anticipated in the promulgation of the system of a regularly broken-stone covering by Mr. Edgeworth, an Irish proprietor, whose treatise on roads, of which the second edition was published in 1817,* contains the results of his experiments on the construction of roads, with some useful rules. He advocated the breaking of the stones to a small size, and their equal distribution over the surface. He also recommended that the interstices should be filled up with small gravel or sharp sand—a practice which, though it was condemned by Macadam, is now adopted by the best surveyors.

Since Macadam's time, the practice of road-making has been greatly improved by the use of the roller for compressing and settling new materials, and of preparing at once a comparatively smooth and hard surface for traffic.

Telford first directed his attention in 1803-4, to the construction of roads. He was employed chiefly in the construction of new roads—hundreds of miles of roads in the Scottish Highlands; also the high road from London to Holyhead and Liverpool, and the great north roads, formed in consequence of the increased communication with Ireland after the Union, and which were excellent models for roads throughout the kingdom. Telford set

* "An Essay on the Construction of Roads and Carriages," 2nd edition, 1817.

out the roads according to the wants of the district through which they were made, as well as with a view to more distant communication; and the acclivities were so laid out, that horses could work with the greatest effect for drawing carriages at rapid rates. As a notable instance of the wonderful improvements that were effected by Telford's engineering skill applied to the laying out of new roads, an old road in Anglesea rose and fell between its extremities, 24 miles apart, through a total vertical height of 3,540 ft.; whilst a new road, laid out by Mr. Telford between the same points, rose and fell only 2,257 ft., or 1,283 ft. less than the undulations of the old road, whilst the new road was more than 2 miles shorter.

The road was formed by a substratum, or rough hand-set pavement, of large stones as a foundation, with sufficient interstices between the stones for drainage. The materials laid on this foundation were, like Macadam's materials, hard and angular, broken into small pieces, decreasing in size towards the top, where they formed a fine hard surface, whereon the carriage wheels could run with but little resistance. Telford's system was afterwards studied by his assistant, Mr. (afterwards Sir John) Macneil.

The pressure of public opinion, acting through more than a century, has resulted in a network of fully 160,000 miles of good carriageable roads in the United Kingdom, according to the following data supplied by Mr. Vignoles:—*

Length of Metalled Roads in 1868-69.

	Length of Road. Miles.	Area. Square miles.	Population. Numbers.
United Kingdom	160,000	122,519	30,621,431
France	100,048	210,460	38,192,064
Prussia	55,818	139,675	23,970,641
Spain	10,886	198,061	15,673,481

The rolling of Macadam or broken-stone roads, though

* Address of the President of the Institution of Civil Engineers, January 11th, 1870.

it seems to have been first applied in 1830, appears to have been but imperfectly appreciated in England until about the year 1843, when, according to Mr. F. A. Paget, the first published recommendation in the English language of horse road-rolling, as a measure of economy, was issued by Sir John Burgoyne.* Road-rolling is now very generally practised, by horse-power or by steam-power.†

The first Act for paving and improving the City of London was passed in 1532. The streets were described, in this simply-worded statute, as "very foul, and full of pits and sloughs, so as to be mighty perilous and noyous, as well for all the king's subjects on horseback, as on foot with carriages" (litters).

Previously to the introduction of the turnpike-road system, the streets of the metropolis and other large towns were paved with rounded boulders, or large irregular pebbles, imported from the sea-shore. They usually stood from 6 to 9 inches in depth for the carriage-way, and about 3 inches deep for the footpaths. Such a road could not be made with a very even surface; the joints were necessarily very wide, and afforded receptacles for filth. The irregularity of the bases of the stones caused a difficulty in securing a solid and equal support; and, under the traffic, ruts and hollows were speedily formed. The boulder pavement was succeeded by a pavement composed of blocks of stone which, though ordinarily of tolerably good quality, and measuring 6 or 8 inches across the surface, were so irregular in shape that even their surfaces did not fit together. They formed a rubble causeway,

* See a paper by Sir John Burgoyne "On Rolling new-made Roads," in the Appendix.

† The history of Horse Road-Rolling and of Steam Road-Rolling, is given by Mr. Frederick A. Paget in his instructive "Report on the Economy of Road Maintenance and Horse-draught through Steam Road-rolling; with Special Reference to the Metropolis, 1870." Addressed to the Metropolitan Board of Works.

in which the stones were but slightly hammer-dressed. Wide joints were made; and far from being dressed square down from the surface, they most frequently only came into contact near the upper edges; and, tapering downwards, their lower ends were narrow and irregular, leaving an insufficient area of flat base to support weight. With such irregular forms, considerable spaces were unavoidably left between the stones, which were filled by the paviours with loose mould, sand, or other soft material, of which the bed or subsoil was composed. Another great deficiency in the construction of the pavement, was caused by inattention to the selection and arrangement of the stones according to size—large and small stones were placed alongside of each other, and, as they acted unequally in their resistance to pressure, they created a continual jolting in wheel-carriages, and, adding percussive action to pressure, became powerful destructive agents. Again, the bed on which the stones were placed, being loose matter, for the most part, was easily converted into mud when water sank through between. It was unavoidably loosened by the paviour's tool, to suit the varying depths and narrow bottoms of the stones, and to fill up the chasms between the stones. The mud was worked up to the surface, and the stones were left unsupported. In consequence of these defects, the surface of the pavement soon became very uneven, and not unfrequently sunk so much as to form hollows, which rendered it not only incommodious but dangerous to horses and carriages.

Such was the system of pavement met with in London fifty years ago. Mr. Telford, in 1824, clearly pointed out the deficiencies of the system; and in his Report (referred to in the foot-note)* he recommended first, a bottoming,

---

* See Mr. Telford's "Report respecting the Street Pavements, &c., of the Parish of St. George's, Hanover Square," printed in Sir Henry Parnell's "Treatise on Roads," p. 348, 2nd Edition.

or foundation, of broken stones, 12 inches deep; second, rectangular paving-stones of granite, worked flat on the face, straight and square on all the sides, so as to joint close, with a base equal to the face, forming, in fact, an ashlar causeway. The dimensions of the stones were recommended to be as follows:—

	Width. Inches.	Depth. Inches.	Length. Inches.
For streets of the 1st class	6 to 7½	10	11 to 13
,, 2nd ,,	5 to 7	9	9 to 12
,, 3rd ,,	4½ to 6	7 to 8	7 to 11

Stones of such dimensions as those recommended by Telford, frequently having a depth of 12 inches, have been generally employed in street-paving. In some instances, they have been laid on concrete, with the joints grouted with lime and sand, to insure a great degree of stability. They have been proved to possess great durability—of which many instances will be adduced—but they have been, for several reasons, generally abandoned in favour of narrower paving-stones, 3 or 4 inches in width, though many secondary streets in London and elsewhere, remain, at this day, paved with 6-inch stones.

Macadam's system was introduced in some streets where the traffic was light, but it did not equal the granite paving.

Pavements formed of blocks of wood appear to have been first employed in Russia, where, according to the testimony of Baron de Bode,* it has been, though rudely fashioned, used for some hundreds of years. After long and repeated trials of various modes of construction, wood pavement consisted, according to the approved method, of hexagonal blocks of fir wood, 6 inches across and 7 inches deep, planted, with the fibre vertical, close to each other, on a sound and level bottom; a boiling mixture of

* "Wood Pavement," by A. B. Blackie, 1842.

pitch and tar was poured over them, and a small quantity of river sand was strewed over the tar. "The fabrication of these blocks," wrote Baron de Bode, "is extremely simple and expeditious. It is accomplished by fastening six strong blades into a strong bottom of cast-iron, and pressing the ready-cut pieces of wood through these six blades by means of a common or hydraulic press. The bottom of the press being open, these cut blocks drop on the floor, completely formed for immediate use. Red fir is considered the best; but none of it must be used when it has blue stripes on its edges, as that is a proof that it is in a state of decay. The blocks must be perfectly dried before they are used, and squeezed as close together as possible between the abutments, one on each side of the street or road, so as to keep the pavement from moving." In Norway, Sweden, Denmark, and Iceland, wood was, at the time of Mr. Blackie's writing, and it may be now, in general use for the pavement of streets and highways.

In the United States, likewise, wood pavement was laid down experimentally—in New York in 1835-6, and about the same time in Philadelphia. In New York, it was laid in three different forms. A hundred yards was laid in

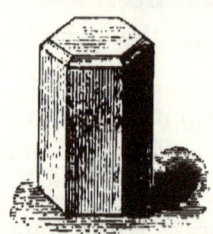

Fig. 5.—Stead's Wood Pavement. Form of Blocks.

Broadway, consisting of hexagonal blocks of pitch-pine, 6 inches across, and 8 inches deep. No pitch or tar was applied to this pavement: it was simply strewed occasionally with gravel or sand for a month after it was laid. It had lain for two years, according to report, without having required any repair; though it appears that very few carts passing over it carried more than half a ton of load, of which the widest wheel did not exceed three inches in width. An equal length of pavement was laid in William Street, a minor thoroughfare, in the end of 1836; the pavement consisted of 6-inch square blocks of pine, 12 inches deep. The third specimen was laid in Mill Street, in the middle of 1837, consisting of the same size and kind of blocks as those laid in William Street, on a foundation of sand beat down very hard. It is stated in Mr. Blackie's pamphlet that the pavement of square blocks was laid on boards—probably in William Street.

Mr. David Stead was the first constructor of wood pavement in England. He patented his system in May, 1838:—consisting of hexagonal blocks of Scotch fir or Norway fir, from 6 to 8 inches across, and from $3\frac{1}{2}$ to 6 inches deep, according to the traffic of the thoroughfare in which they were to be laid. Each block was of the form shown in Fig. 5, chamfered at the upper edges. The ground having been well beaten and levelled, it was covered with three inches of gravel, upon which the blocks were placed, and which was designed to carry away the water which might penetrate below the surface. The pavement, when completed, looked substantial, and presented the appearance shown in Fig. 6. When the blocks were grooved across, they appeared together as in Fig. 7. Mr. Stead's pavement was, in several instances, laid on a bed of concrete. In Manchester, where it was thus laid, in front of the Royal Infirmary, the concrete bed was three inches deep, and was composed of three parts of small broken stones, $\frac{3}{4}$ inch in diameter, flushed with Ardwick lime and Roman cement. The lime was mixed with sand in the proportion of one to two; and the cement as one to twenty. The concrete was laid upon a hard, well-beaten clay substratum.

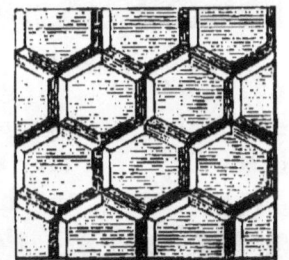

Fig. 6.
Stead's Wood Pavement, 1838.

Mr. Stead also laid pavements experimentally, consisting of round blocks of wood—sections of trees—placed vertically, and laid together as in Fig. 8. The interspaces were filled with sifted gravel or sharp sand.

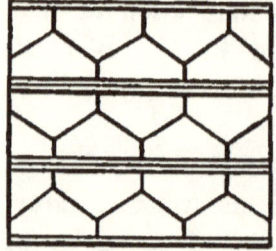
Fig. 7.
Stead's Wood Pavement, 1838.

The first example of wood-paving in London, was laid in the Old Bailey, in 1839, on Stead's system. It was laid haphazard on the bed of the roadway. The pavement did not wear well; the blocks settled down irregularly in the unprepared foundation. At the end of three years and two months, in 1842, the pavement was lifted, and removed to pave the yard of the Sessions House; there it decayed, and a large crop of fungi appeared in the places not touched by the traffic.

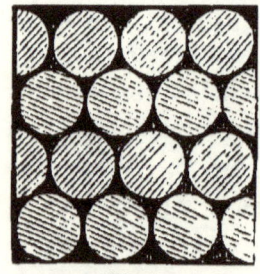
Fig. 8.
Stead's Wood Pavement. Round Blocks.

Mr. Stead's system of wood paving was laid in several other localities in London about the same time as the piece which was laid in the Old Bailey, and also in Woolwich Dockyard. It was laid also in Salford, Liverpool, and Leeds.

Shortly after Mr. Stead's attempt, during the period from 1840 to 1843, seven other wood pavements, of various design, were laid in the City; but they did not last, for the most part, more than three or four years. One of these was the invention of the Count de Lisle, patented in the name of Hodgson, in December, 1839; the invention was acquired by the Metropolitan Wood Pavement Company. The formation of the blocks was called the "Stereotomy of the Cube." The upper and under surfaces of the blocks, Fig. 9, are cut diagonally to the direction of the grain,

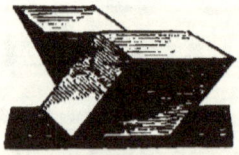
Fig. 9.—De Lisle's Wood Pavement. Form of Blocks, 1839.

forming parallelopipeds, which are placed alternately in reversed positions, and when put together present a pavement having the appearance of Fig. 10. In each block, two holes are cut on each side to receive dowels or trenails, designed to lock the blocks together.

Fig. 10.—De Lisle's Wood Pavement, 1839.

Mr. Carey's wood pavement, patented in 1839, was one of the earliest pavements that were tried, and it proved to be the best at the time. It was first laid in the City, in the Poultry, in 1841, where it lasted six years; and it was shortly afterwards laid in many other streets. It consisted of blocks of wood 6 or 7 inches wide, from 12 to 14 inches in length, and 8 inches deep, shown in side elevation, Fig. 11. The four-sided blocks of wood were of wedge-form, in and out, sidewise and endwise vertically, so as to form salient and re-entering angles, and to interlock on all the four sides, each block with its neighbour, when laid. It was anticipated that, by this arrangement, each block would receive support from its neighbours, and would be prevented from shifting or settling from its position, since the pressure of the load that was to come upon each block in succession would be distributed and dispersed over the neighbouring blocks. Later experience has demonstrated two things:—that lateral support of this kind was not required; and that, following the experience of stone-set paving, the wood blocks of narrower dimensions answered better, and, with suitable interspaces, afforded the necessary foothold for horses.

Fig. 11 — Carey's Wood Pavement. Vertical Section, 1839.

Asphalte, a natural, brittle compound of bitumen and

limestone, found in volcanic districts, was introduced from France, for foot-pavements, in 1836. It has, since that time, been extensively employed in the City of London for the pavements of carriage-ways.

In France, the art of the construction of roads, a hundred years ago, was far in advance of English practice. Previously to 1775, the causeway was generally 18 feet wide, with a depth of 18 inches at the middle and 12 inches at the sides, according to the profile, Fig. 12. Stones were laid

Fig. 12.—Section of French Roads. Previous to 1775.

flat, by hand, in two or more layers, on the bottom of the excavation; on this foundation, a layer of small stones was placed and beaten down, and the surface of the road was formed and completed with a finishing coat of stones broken smaller than those immediately beneath. As the roads were, down to the year 1764, maintained by statute labour, with which the reparations could only be conducted in the spring and the autumn of each year, it was necessary to make the thickness of the roads as much as 18 inches, that they might endure during the intervals between repairs. With less depth, they would have been cut through and totally destroyed by the deep ruts which were formed in six months.

The suppression of statute labour (*la corvée*), in 1764, was the occasion of a reformation in the design of causeways, whereby the depth was reduced to such dimensions as were simply strong enough for resisting the weight of the heaviest vehicles. The depth was reduced to a uniform dimension of 9 or 10 inches from side to side, and the cost was diminished more than one half. Writing in 1775, M. Trésaguet, engineer-in-chief of the generality of Limoges, stated that roads constructed on the improved plan lasted for ten years, under a system of constant maintenance, and that they were in as good condition as when first constructed. The section of these roads, as elaborated by M. Trésaguet, is shown in Fig. 13. The form of the bottom is

Fig. 13.—Section of French Roads, elaborated by M. Trésaguet. 1775.

a parallel to the surface, at a depth of 10 inches below it. Large boulder stones are laid at each side. The first bed consisted of rubble stones laid compactly edgewise, and beaten to an even surface. A second bed, of smaller stones, was laid by hand upon the first bed. Finally, the finishing layer, of small broken stones, broken by hand to the size of walnuts, was spread with a shovel. Great care was taken in the selection of stone of the hardest quality for the upper surface. The rise of the causeway was 6 inches in the width of 18 feet, or 1 in 36.

Trésaguet's method, here illustrated, was generally adopted by French engineers in the beginning of the present century; although, on soft ground, they placed a layer of flat stones on their sides under the rubble work. In this case, the thickness was brought up to 20 inches. The rise of the causeway was as much as 1-25th, and often equal to 1-20th of the width.

But, if the design was good, the maintenance was bad. Large and unbroken stones were thrown into the holes and ruts, and neither mud nor dust was removed. About the year 1820, the system of Mr. Macadam attracted some attention in France; and the peculiar virtue of angular broken stone in closing and consolidating the surface was recognised. About the year 1830, it is said, the system of Macadam was officially adopted in France for the construction of roads; and M. Dumas, engineer-in-chief of the Ponts et Chaussées, writing in 1843,* stated that the system of Macadam was generally adopted in France, and that the roads were maintained, by continuous and watchful attention in cleansing the roads and with constant repair, in good condition—realising his motto, "The maximum of beauty." But the employment of rollers for the preliminary consolidation and finishing of the road, has been an essential feature in their construction and their maintenance; for it has long been held in France that a road unrolled is only half finished. It appears, according to Mr. F. A. Paget, that the horse-roller was introduced in France in 1833. At all events, in 1834, M. Polonceau, struck by the viciousness of the mode of aggregating or rolling the material of the road by the action of wheels, proposed, in the first place, to consolidate the bottom by a 6-ton roller, and to roll the material in successive layers consecutively, and thus to complete in a few hours what might, in the ordinary course of wheel-rolling, require many months to perform.

* "Annales des Ponts et Chaussées," 1843; tome 5, page 348.

PART I.

CONSTRUCTION OF ROADS.

BY HENRY LAW, C.E.

CHAPTER I.

EXPLORATION FOR ROADS.

THIS part of the work is confined to the art of constructing common roads, in situations where none previously existed, and to the repair of those already made. Before entering into the details of their construction, it is desirable to go into the subject of the *exploration* for roads, or the manner in which a person should proceed in exploring a tract of country, for the purpose of determining the best course for a road, and the principles which should guide him in his final selection of the same.

Suppose that it is desired to form a road between two distant towns, A and B, Fig. 14, and for the present neglect

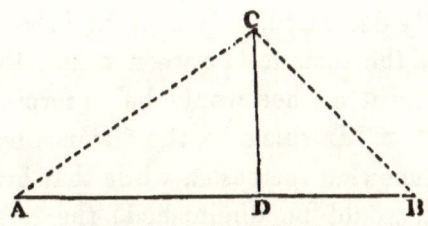

Fig. 14.—Laying out a Road.

altogether the consideration of the physical features of the intervening country; assuming that it is equally favourable, whatever line is selected. Now, at first sight, it would appear that, under such circumstances, a perfectly straight line drawn from one town to the other, would be the best that could be chosen. But on a more careful examination of the locality, it may be found that there is a third town, c, situated somewhat on one side of the straight line drawn from A to B; and, although the primary object is to connect the two latter, it may, nevertheless, be considerably better if the whole of the three towns were put into mutual connection with each other. Now this may be effected in three different ways; any one of which might, under certain circumstances, be the best. In the first place, a straight road might, as originally suggested, be formed from A to B, and, in a similar manner, two other straight roads from A to c, and from B to c. This would be the most perfect way of effecting the object in view, the distance between any two of the towns being reduced to the least possible length. It would, however, be attended with considerable expense, and it would be requisite to construct a much greater length of road than according to the second plan, which would be to form, as before, a straight road from A to B, and from c to construct a road which should join the former at a point D, so as to be perpendicular to it; the traffic between A or B and c, would proceed to the point D, and then turn off to c. By this arrangement, while the length of the roads would be very materially decreased, only a slight increase would be occasioned in the distance between c and the other two towns. The third method would be to form only the two roads A c and c B. In this case, the distance between A and B would be somewhat increased, while that between A and c, or B and c, would be diminished; the total length of road to be constructed would also be lessened.

As a general rule, it may be taken that the last of these methods is the best, and most convenient for the public;

that is to say, if the physical character of the country does not determine the course of the road, it will generally be found best not to adopt a perfectly straight line, but to vary the line so as to pass through the principal towns near its general course. The public may thus be conveyed from town to town with greater facility and less expense than if the straight line were adopted, and the towns were to communicate with it by means of branch roads. On the first system, vehicles established to convey passengers or goods between the two terminal towns, would pass through all those which were intermediate; whilst, if the straight line and branch-road system were adopted, a system of branch coaches would be required for meeting the coaches on the main line.

In laying out a road in an old country, in which the position of the several towns, or other centres of industry, requiring road accommodation, is already determined, there is less liberty for the selection of the line of road than in a new country, where the only object is to establish the easiest and best road between two distant stations. In the first case, the positions of the towns, and other inhabited districts situated near the intended road, are to be taken into consideration, and the course of the road may, to a certain extent, be controlled thereby; whilst, in the second case, the physical character of the country would alone be investigated, and it alone would constitute the basis for the selection of a new route.

Whichever of these two cases may be dealt with, in the selection and adoption of the line of road between two points, a careful examination of the physical character of the country should be made, and the line of the route determined in accordance with physical conditions.

One of the first points which attract notice in making an examination of an ordinary tract of country, is the unevenness or undulation of its surface; but if the observation be extended a little further, one general principle of conformation is perceived even in the most irregular countries. The country is intersected in various directions by rivers, increasing in size as they approach their point of discharge; towards these main rivers, lesser rivers approach on both sides, running right and left through the country; and into these, again, enter still smaller streams and brooks. Furthermore, the ground falls in every direction towards the natural watercourses, forming ridges, more or less elevated, running between them, and separating from each other the districts drained by the streams.

It is the first business of a person, engaged in laying down a line of road, to make himself thoroughly acquainted with the features of the country; he should possess himself of a plan or map, showing accurately the course of all the rivers and principal watercourses, and upon this he should further mark the lines of greatest elevation, or the ridges separating the several valleys through which they flow. It is also of peculiar service when the plan contains contour lines showing the comparative levels of any two points, and the rates of declivity of every portion of the country's surface. The system of showing upon plans the levels of the ground by means of *contour-lines*, is one of much utility, not only in the selection of roads and other lines of communication, but also for settling the lines of the drainage of towns, as well as of their water-supply, and of the drainage and irrigation of lands, and for many other purposes. A contour-plan of the City of London * (Fig. 15) illustrates the application of the system of contour levels. It will be observed that, upon this plan, there are a number of fine lines traversing its surface in various directions, and, where they approach the borders of the map, having figures written against them: these lines are

* This plan is taken from a Report on the Health of Towns, and is made from levels taken from Mr. Butler Williams.

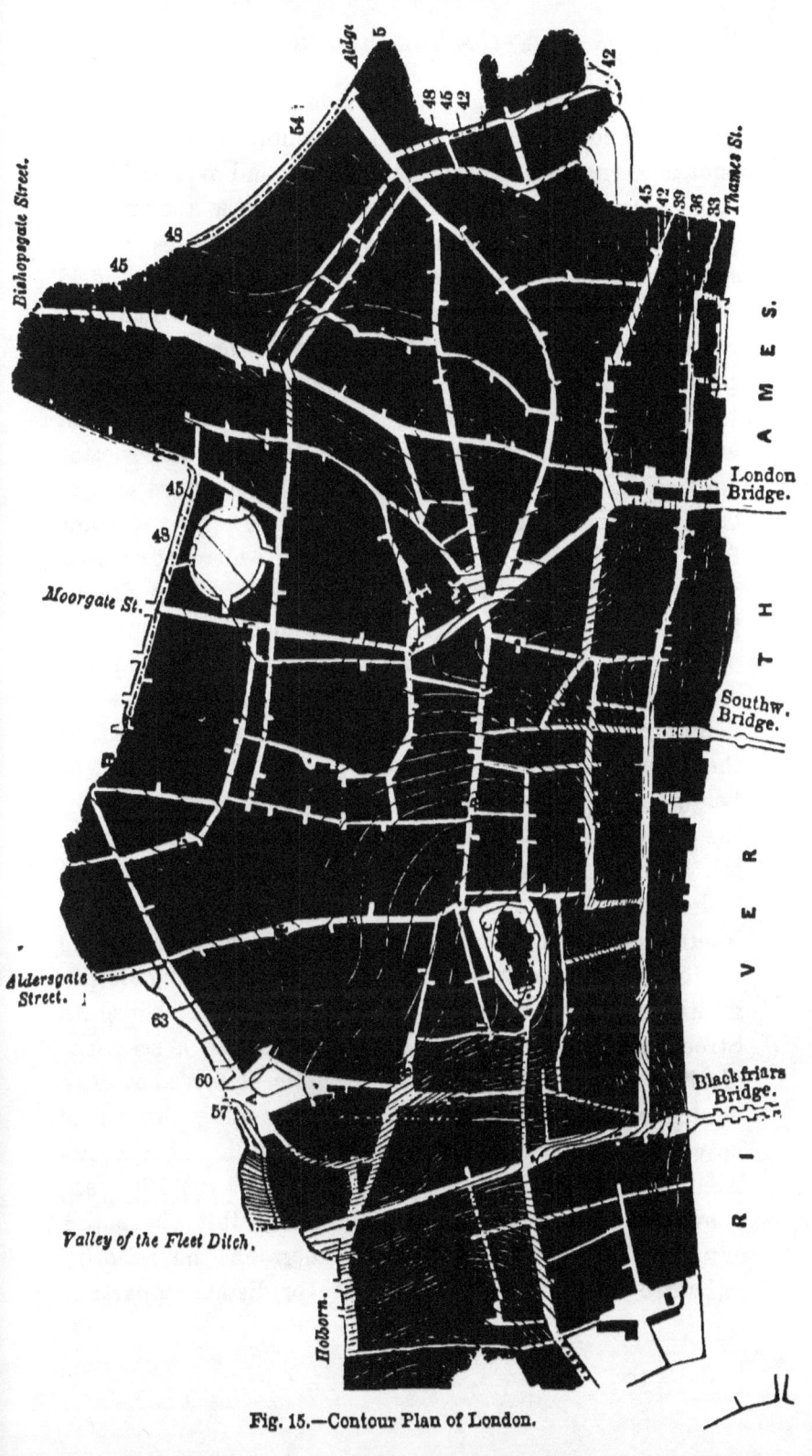

Fig. 15.—Contour Plan of London.

termed *contour-lines*, and they denote that the level of the ground is identical throughout the whole of their course: that is to say, that every part of the ground over which the line passes, is at a certain height above a known fixed point, the height being indicated by the figures written against the line. At the point A, for example, in Smithfield Market, a line with the figures 57 is attached, which indicates that the ground at that spot is 57 feet above some point to which all the levels are referred. If the course of the line be traced, it is found that it cuts Newgate Street at the point B, passes thence to the bottom of Paternoster Row at the point I, through St. Paul's Churchyard at c, to Cheapside at D. It then curves round towards the point from which it first started, and crosses Aldersgate Street twice, at E and F; and, after intersecting Fore Street, Cripplegate, in the point G, it again meets the boundary of the City at H. It is thus shown that, tracing the course of this line, each of those points stands at the same height, namely, 57 feet above a certain fixed point, termed the *datum*. This point is, in the present instance, 10 feet below the top of the cap-stone at the foot of the step, on the east side of Blackfriars Bridge. Each interval between the lines in Fig. 8, indicates a difference of level of 18 inches; and by counting the number of these lines which intersect a street or road within any given distance, the rise or fall in the street is at once ascertained by simple multiplication. Thus, looking at the line of Bishopsgate Street, near the north end, the contour-line 45 is seen, indicating that that point in the street is 45 feet above the datum, and nine lines are found intersecting the street between that point and the top of Cornhill. It is calculated, therefore, that this point is ($1.5 \times 9 =$) 13.5 feet above the other end of the street, or 58.5 feet above the datum. The rate of inclination of the ground may also be estimated by the relative proximity or distance apart of these lines. Thus, on the northern side of the City, where the ground is comparatively level, the lines are far apart; whereas, on the side next the Thames, and again on each side of the line of Farringdon Street, which marks the course of the valley of the old river Fleet, where the surface is very hilly, the contour lines lie close together.

The plan, Fig. 16, shows an imaginary tract of country, to illustrate more clearly the mode of showing by means of contour-lines, the physical features of a country. The hatched line, E F G H I, is supposed to be an elevated ridge, encircling the valley shown in the plan; the fine black lines are contour-lines, indicating that the ground over which they pass is at the altitude above some known mark expressed by the figures written against them in the margin. It will be observed that these lines, by their greater or less distance, have the effect of shading, and make apparent to the eye, the undulations and irregularities in the surface of the country.

In laying out a line of road, there are three cases which may have to be treated, and each of these is exemplified in the plan, Fig. 16. First, the two places to be connected, as the towns A and B on the plan, may be both situated in the same valley, and upon the same side of it; that is, that they are not separated from each other by the main stream which drains the valley. This is the simplest case. Secondly, although both in the same valley, the two places may be on the opposite sides of the valley, as at A and C, being separated by the main river. Thirdly, they may be situated in different valleys, separated by an intervening ridge of ground more or less elevated, as at A and D. In laying out an extensive line of road, it frequently happens that all these cases have to be dealt with: frequently, perhaps, during its course.

The most perfect road is that of which the course is perfectly straight, and the surface perfectly level; and, all

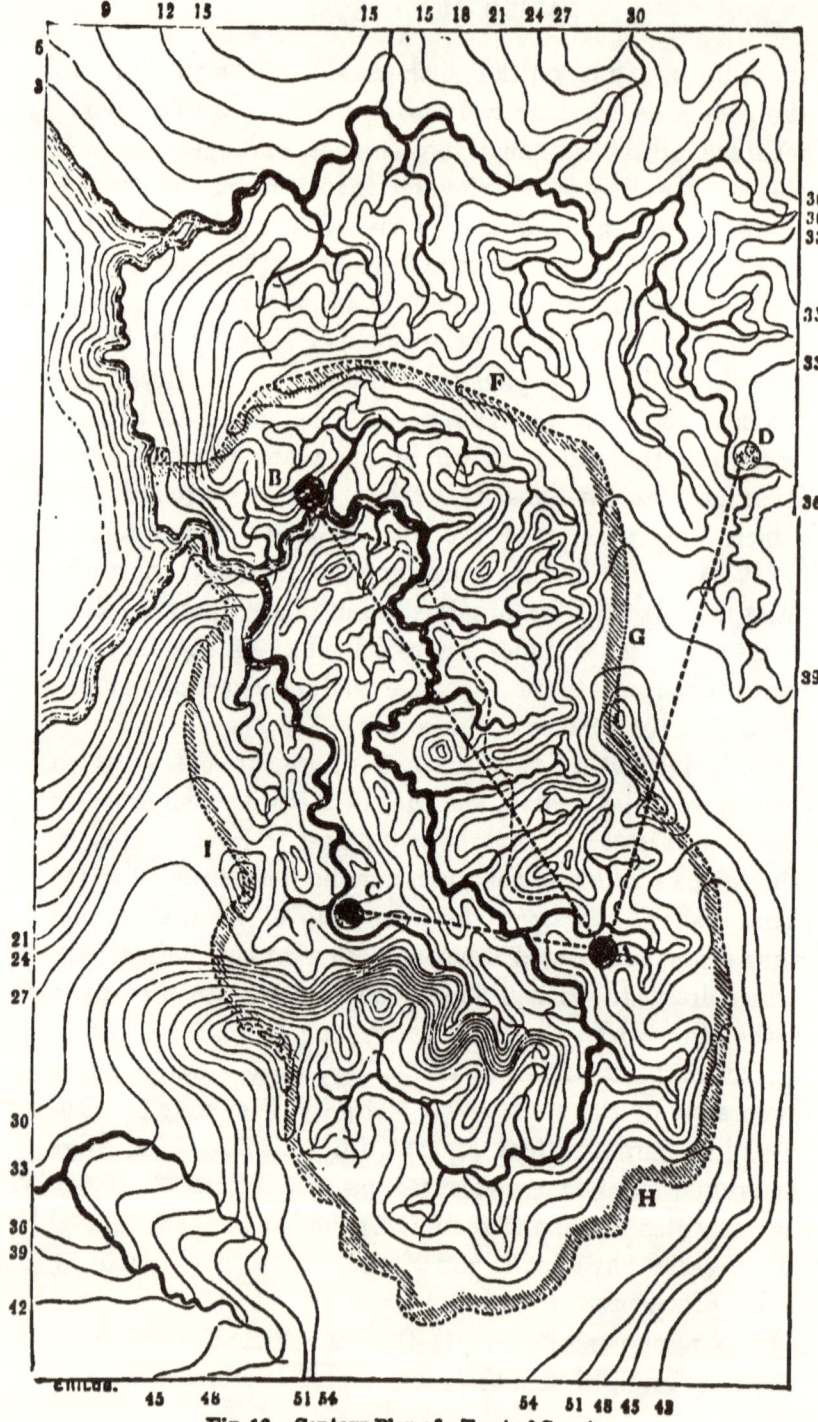

Fig. 16.—Contour Plan of a Tract of Country.

other things being the same, that is the best road which answers nearest to this description.

Now, in the first case:—That of the two towns situated on the same side of the main valley, there are two methods which may be pursued in forming a communication between them. A road following the direct line between them, shown by the thick dotted line A B may be made; or, a line may be adopted which should gradually and equally incline from one town to the other, supposing them to be at different levels, or which should keep, if they are on the same level, at that level throughout its entire course, following all the sinuosities and curves which the irregular formation of the country might render necessary for the fulfilment of these conditions. According to the first method, a level or a uniformly-inclined road might be made from one to the other, forming embankments and cuttings where necessary; or these expensive works might be avoided, and the surface of the road made to conform to that of the country. Now, of all these, the best is the straight and uniformly-inclined, or the level road, although at the same time it is the most expensive. If the importance of the traffic passing between the places is not sufficient to warrant so great an outlay, it will become a matter of consideration whether the course of the road should be kept straight, its surface being made to undulate with the natural face of the country; or whether, a level or equally-inclined line being adopted, the course of the road should be made to deviate from the direct line, and follow the winding course which such a condition is supposed to necessitate.

In the second case, that of two places situated on opposite sides of the same valley, there is, in like manner, the choice of a perfectly straight line to connect them, which would probably require a heavy embankment if the road were kept level; or steep inclines, if it followed the surface of the country; or, by winding the road, it may be carried across the valley at a higher point, where, if the level road were taken, the embankment would not be so high, or, if kept on the surface, the inclination would be reduced.

In the third case, there is, in like manner, the alternative of carrying the road across the intervening ridge in a perfectly straight line, or of deviating it to the right or the left, and crossing at a point where the ridge is less elevated.

The proper determination of the question, which of these courses is the best under certain circumstances, involves a consideration of the comparative advantages and disadvantages of inclines and curves. What additional increase in the length of a road would be equivalent to a given inclined plane upon it; or, conversely, what inclination might be given to a road, as an equivalent to a given decrease in its length? To satisfy this question, it is requisite to know the comparative force required to draw different vehicles with given loads upon level and upon variously-inclined roads:—a subject which is treated in the next chapter.

In laying out a new line of road, the first proceeding is usually, after a general examination of the country, to lay down one or more lines upon the best map which can be procured. On a contour-map of the district, this proceeding is greatly facilitated. The next step is to make an accurate survey of the lands through which the several lines sketched out pass, which should be plotted to such a scale as will admit of the smallest features being shown with sufficient accuracy and distinctness. A scale of ten chains to the inch, for the open country, with enlarged plans of towns and villages upon a scale of three chains to the inch, is generally found to be sufficient. Careful levels should also be taken along the course of each line; and at suitable distances, depending upon the nature of the country, lines of levels should be taken at right angles to the original

PLANS OF ROADS.

line. In taking these levels, the heights of all existing roads, rivers, streams, or canals should be noted; *bench-marks* should be left at least every half-mile, that is, marks made on any fixed object, such as a gate-post, or the side of a house or barn, the exact height of which is ascertained, and registered in the level-book. The bench-marks are useful in case of deviations being made in any portion of the lines, for the levels may be taken direct from the bench-marks, thus obviating the necessity of again levelling other parts of the line. A section should be formed from the levels, to the same horizontal scale as the general plan, with such a vertical scale as will show with distinctness the inequalities of the ground. If the horizontal scale is ten chains to the inch, the vertical scale may be 20 feet to the inch.

A plan of this kind is exemplified in Fig. 17, plotted to a scale of ten chains to the inch, showing a district through which a road is to be constructed. One line is shown running nearly straight across the plan, together with a deviation therefrom, which, although of greater length, would run on more favourable ground. The sections, Figs. 18 and 19, show the levels of the surface of the ground on the straight line, and on the deviation from it respectively. The required information is given on the plans, for enabling the engineer to lay down the course of the road, and to arrange the position and dimensions of the culverts, bridges, and other works necessary in its construction.

It is shown in Fig. 17 that the straight line crosses a stream at B, and the river twice at C and D; and also that it must pass from D to E, over a swamp or morass of such a nature that, if a solid embankment be formed, it is probable that a very large quantity of earth would be absorbed beyond what is indicated in the section. It would, in addition, be necessary to form bridges with several capacious openings at the points where the intended road would

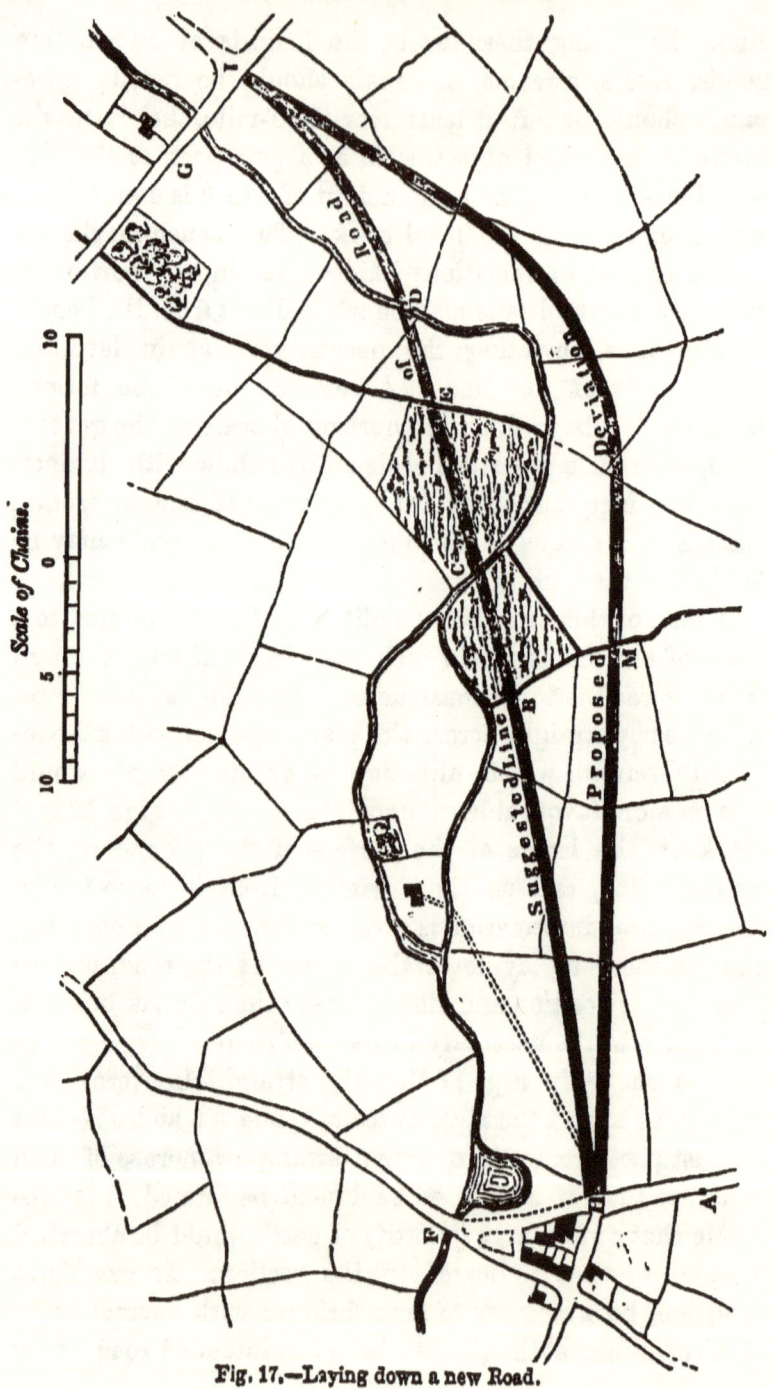

Fig. 17.—Laying down a new Road.

cross the river, since the river would be liable to be flooded. Such disadvantages attending the more obvious route, would induce the engineer to sketch out some other line, by which they would be avoided. He would have the levels taken, with other needful information, to enable him to choose between the two routes.

The manner in which the sections should be drawn, and the nature of the information to be given upon them, are exemplified in Figs. 18 and 19. In addition, data of the following character should be obtained, and should be entered either in the survey field-book, or in the level-book.

At the point B, fig. 10, the line crosses a stream 8 feet in width and 1 foot deep; in flood, this stream brings down a considerable quantity of water.

At the point C on the section, the river is much narrower and is not so deep as at other places, in consequence of a great portion of its waters finding a passage through the marshy ground on either side. Its width is 16 feet, and its depth 2 feet; the velocity of the current is 95 feet per minute; the height of its surface at the present time is 30·10 feet above the datum; and the angle of skew which the course of the stream makes with the line of the road is 62 degrees.

At the point D the river is 27 feet wide, and 2½ feet in depth; its velocity 87 feet per minute; the height of its surface above the datum 29·96 feet; and the angle of skew 49 degrees.

The ground from B to E is of a very soft, boggy nature, and full of water.

The height to which the river has risen during the highest flood known, at the bridge at F on the plan, is 35 feet above the datum; the water-way at that time was 90 feet, and the sectional area of the opening through which the water then flowed was 550 square feet. The same flood at the lower bridge, at G on the plan, was 35·3 feet above the datum; the water-way was 102 feet, and the sectional area nearly 600 square feet.

The deviation-line only crosses one stream, at M, on the plan and the section. The width of this stream at present is 15 feet, and its depth 18 inches; but in times of flood it rises to the same height as the river, and brings down a large body of water. The height of its surface at present above the datum is 31·25 feet, and the angle which its course makes with the line of road is 35 degrees.

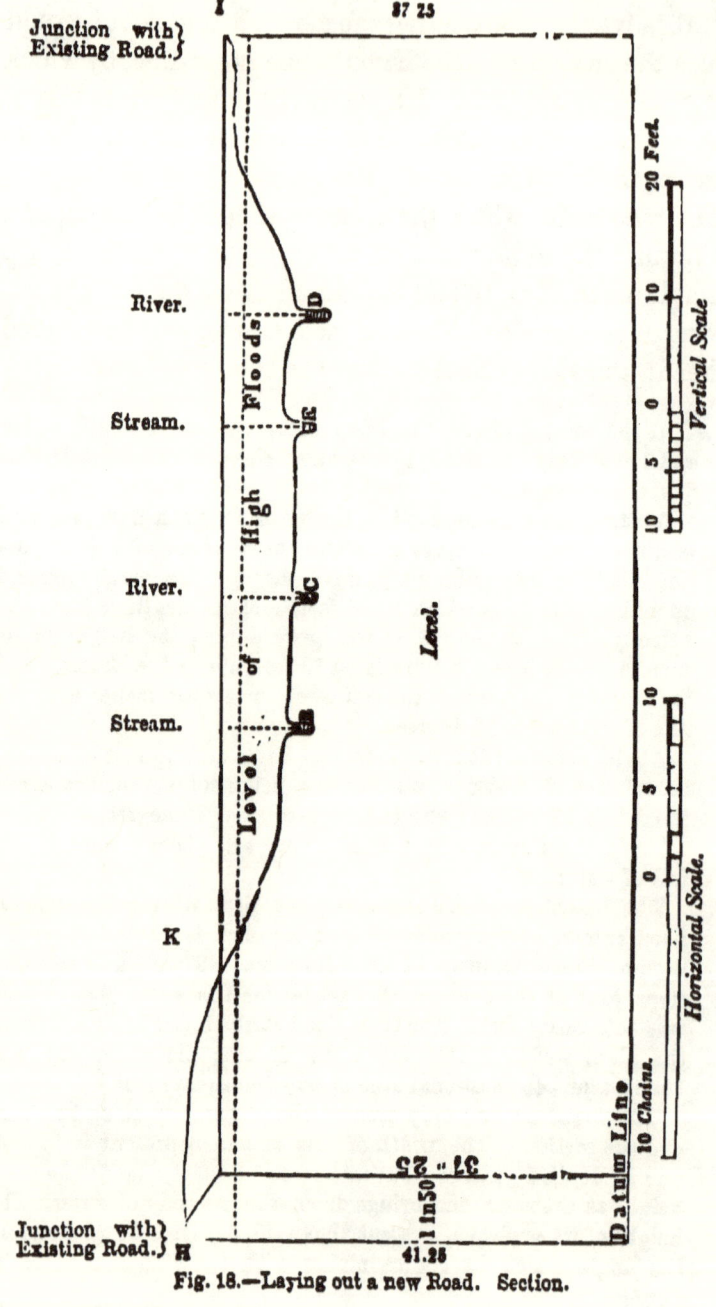

Fig. 18.—Laying out a new Road. Section.

SECTIONAL PLANS OF ROADS. 35

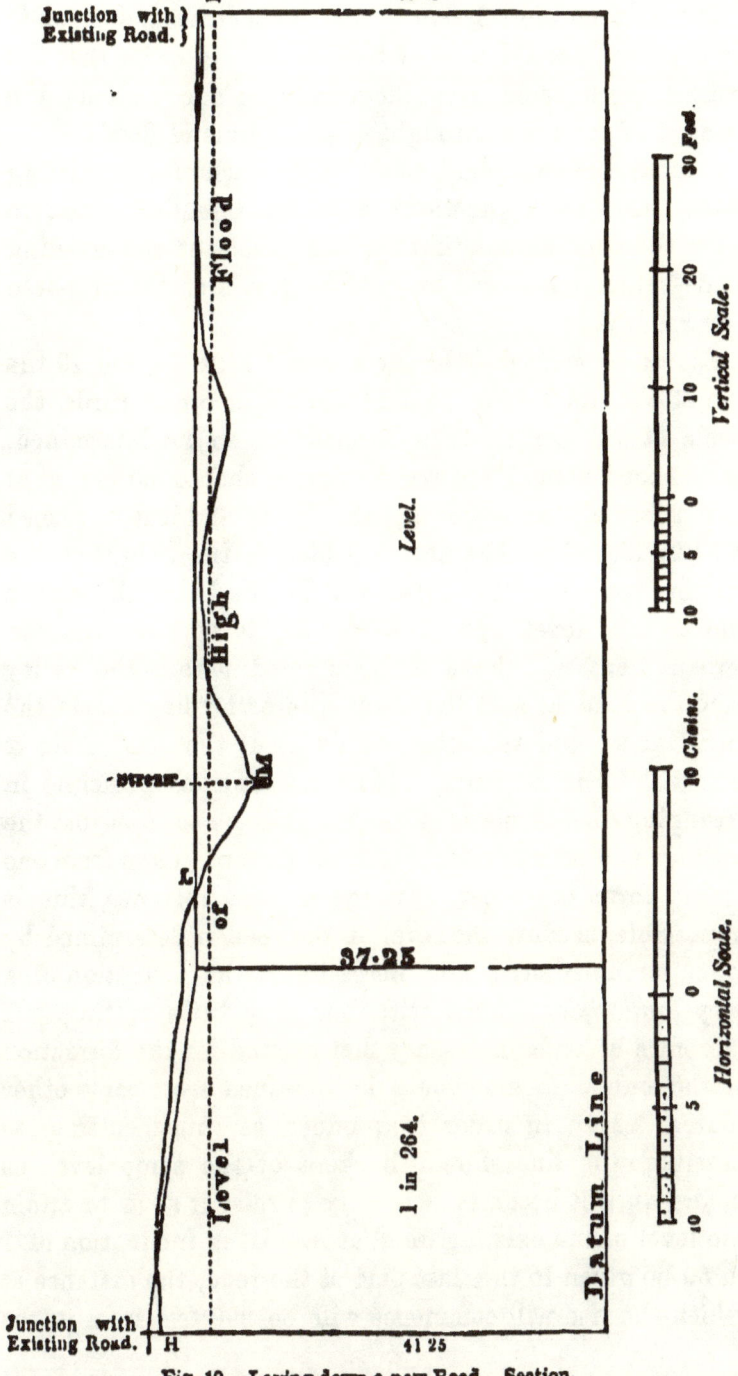

Fig. 19.—Laying down a new Road. Section.

The information relative to the rivers crossed, such as is given above, should always be obtained, in order that the bridges constructed over them may be adequate for the passage of the water brought down in time of floods.

A cross section should be taken of each of the existing roads, near their junctions with the intended road; to show to what extent, if any, the levels of the existing roads might be altered to suit the levels of the proposed new road.

Laying out a Road.—On the sections Figs. 18 and 19 the line of the road is to be laid down; in other words, the levels at which it shall be formed are to be determined. As the road should always be dry, it should be placed at least a foot above the level of the flood; and if it be placed at 37·25 feet above the datum, which is the height of the existing road at I, this object will be effected. Drawing a line at this level upon the section, it appears that an embankment will have to be formed across the valley from the road at I, to the point where the line meets the round at K; and that the remainder of the road from K to H will be in a cutting. Now, the obvious principle in arranging the levels of a road, would be so to adjust the cuttings and embankments that the ground taken from one should form the other. In the present instance, this is impossible, because the level of the road is determined by other circumstances, and necessitates the formation of a very long embankment with but very little cutting. It therefore becomes necessary that ground for the formation of the embankments should be obtained from some other source. But, in order to produce as much cutting as possible, the line should be kept at the same level as before until it becomes necessary to raise it so as to attain the level of the existing road at H. If an inclination of 1 in 50 be given to this last part of the road, the distance at which the rise will commence will be 200 feet from H, the difference of level being 4 feet. There is therefore to be added to the other disadvantages already mentioned, as belonging to the straight line of road, that of the formation of a large embankment, with the necessity for making an excavation in some other place, to supply the earth for that purpose.

Examine the section of the deviation-line, and see what improvement can be thereby effected. The level of the lowest portion of the road must, as before, be placed 37·25 feet above the datum; and if a line be drawn at that level on the section, Fig. 19, it will be found that the quantity of embankment is very much reduced, compared with what would be required for the straight course, and that there is now no difficulty in adjusting the cutting between н and L, so as exactly to afford the amount of filling required. A few trials will show that, if the line be kept at the same level until within sixteen chains of the point н, and then carried up at a regular inclination, this object will be effected, and that the quantities of cutting and embankment will be very nearly equal. The deviation-line is, therefore, the line which the engineer would select as the better of the two. Having made his selection, he would proceed to mark the course of the road on the ground, by driving stakes into the ground, on its centre line, at intervals of one chain-length, or 66 feet. In the next place, he would take very careful levels of the ground at every one of these points, and at any intermediate point, where an undulation or change of level occurred; and wherever the level of the ground varied to any extent in a direction at right angles with the course of the road, he would take levels from which he would make transverse or cross sections of the ground.

From these levels a working section should be made, to a horizontal scale of not less than five chains to the inch, and a vertical scale of 20 feet to the inch. A portion of the section plotted to these scales is shown in Fig. 13; the level of the surface of the ground above the datum, at every chain-length, at the points where stakes have been driven into the ground, should be figured-in on the section, as shown in the column A, and the depth of cutting or height of embankment, at the same points, should be given in another column, B. The entries in this last column are obtained by taking the difference between the level of the surface of the ground and the level of the road. It will be observed that, upon the section, there are two parallel lines drawn as representing the line of road; the upper line is intended to represent the upper surface of the road when finished, while the lower thick line represents what is termed the *formation-surface*, or the level to which the surface of the ground is to be formed, to receive the foundation of the road. In the section, the formation-surface is shown 15 inches below the finished surface of the road; the difference of level is therefore the thickness of the road itself. All the dimensions on the section are understood to refer to the formation-level; and the height of the latter above the datum should be figured-in wherever a change in its rate of inclination takes place, and should be marked by a stronger vertical line, as shown at c.

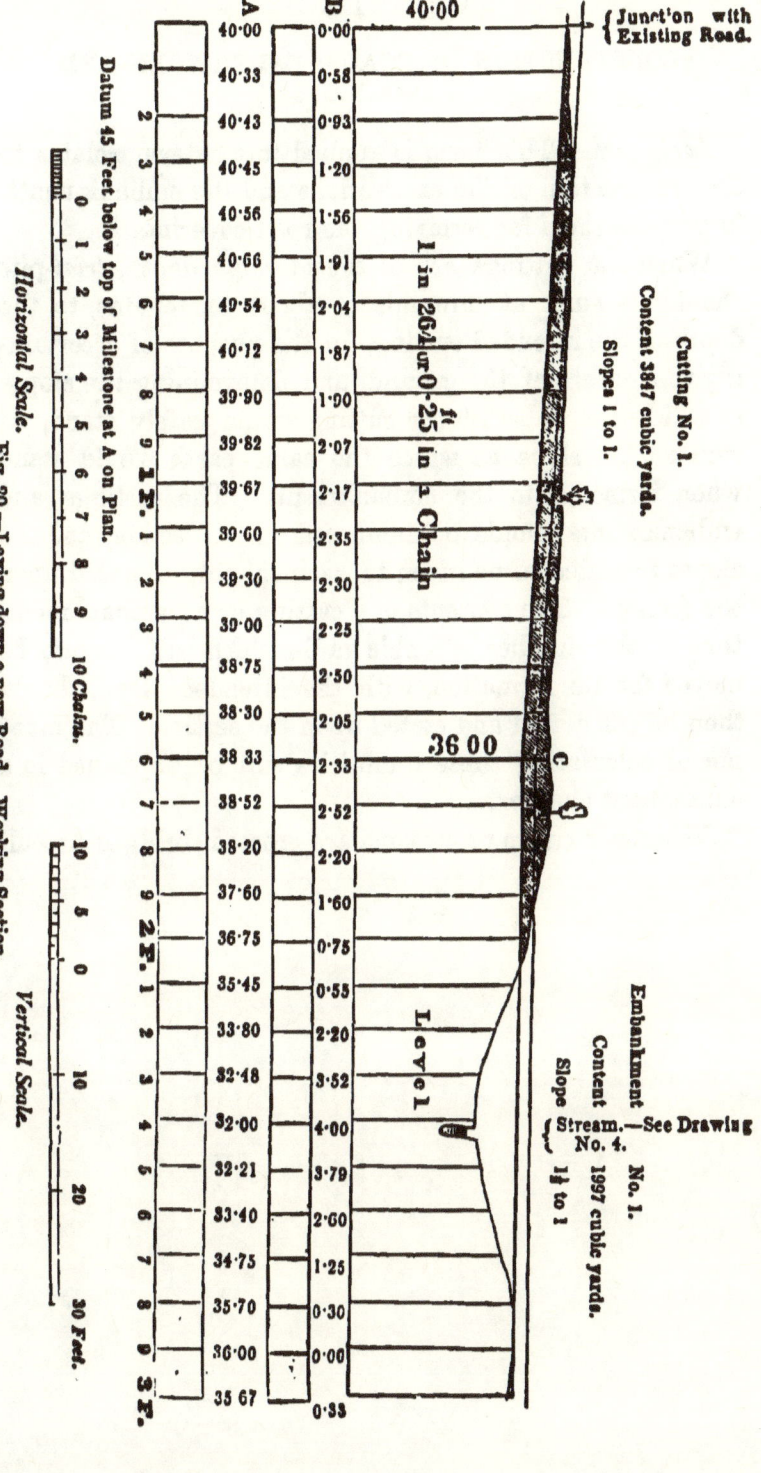

Fig. 20.—Laying down a new Road. Working Section.

CHAPTER II.

CONSTRUCTION OF ROADS: EARTHWORK AND DRAINAGE.

Earthwork.—This term is applied to whatever relates to the construction of the excavations and the embankments, to prepare them for receiving the road-covering.

When the cuttings are of considerable depth, *trial*-pits should be sunk at intervals of about ten chains, to the depth of the intended cutting, for the purpose of ascertaining the nature of the ground, and determining the slopes at which the sides of the cutting would safely stand, as well as the slope at which the same earth would stand when formed into the embankments. The cuttings and embankments should be numbered on the section, and the slopes intended to be given to each should be stated upon the section. The contents of a cutting or an embankment, that is, the number of cubic yards which will have to be moved for its formation, with the intended slope, should then be calculated and stated upon the section. The manner of calculating these quantities will be explained in a subsequent chapter.

Wherever rivers or streams are crossed, bridges or cul-

EARTHWORK.

verts must be introduced; detail drawings of these should be prepared, and reference should be made to them on the working section.

A working plan should be constructed, on the same horizontal scale as the section, upon which the positions of the centre stakes should be shown; and on this plan the road should be drawn to its correct width at the upper surface, with other lines showing the feet of the slopes. The stakes should be numbered consecutively on the plan, to facilitate reference to any part of the line, and the width of land required at every stake should be calculated in the manner about to be described, and entered in a table, from which the width of land required for the purpose of the road may be ascertained at every chain. Suppose that, in the present instance, the finished width of the road itself is to be 20 ft., and that an additional 6 ft. will be required on each side for the ditch and bank, the net width of the road without any slopes, or where the road is on the same level as the ground, would be 26 ft.; and it may be observed in the following table, wherever there is no cutting or embankments (as at stakes Nos. 1 and 30), this is the width given in the fourth column. To find the heights at the other stakes, the product of the height of embankment or depth of cutting (as the case may be) by the ratio of the slope is to be added to the net width, 26 ft. Thus, in the first cutting, the ratio of the slopes being, as stated on the section, 1 to 1, there is simply to add the depths of the cutting at each stake to 26 ft., and the numbers given in the fourth column are obtained. After the 21st stake, the cutting terminates, and the ratio of the slopes then becomes $1\frac{1}{2}$ to 1, and an addition of one and a half times the height of the embankment is to be made to the normal width, 26 ft., to give the remaining values in the fourth column of the table.

EARTHWORK AND DRAINAGE.

TABLE No. 2.—DEPTH, HEIGHT, AND WIDTH OF CUTTINGS AND FILLINGS.

No. of stake on the plan.	Depth of cutting.	Height of embankment.	Distance of side fence from centre line.	No. of stake on the plan.	Depth of cutting.	Height of embankment.	Distance of side fence from centre line.
	Feet.	Feet.	Feet.		Feet.	Feet.	Feet.
1	0·00	—	26·0	17	2·33	—	28·3
2	0·58	—	26·6	18	2·52	—	28·5
3	0·93	—	26·9	19	2·20	—	28·2
4	1·20	—	27·2	20	1·60	—	27·6
5	1·56	—	27·6	21	0·75	—	26·8
6	1·91	—	27·9	22	—	0·55	26·8*
7	2·04	—	28·0	23	—	2·20	29·3
8	1·87	—	27·9	24	—	3·52	31·3
9	1·90	—	27·9	25	—	4·00	32·0
10	2·07	—	28·1	26	—	3·79	31·7
11	2·17	—	28·2	27	—	2·60	29·9
12	2·35	—	28·4	28	—	1·25	27·9
13	2·30	—	28·3	29	—	0·30	26·5
14	2·25	—	28·3	30	—	0·00	26·0
15	2·50	—	28·5	31	—	0·33	26·5
16	2·05	—	28·1				

After ascertaining the widths as shown in the table No. 2, the next operation is to set out the widths on the ground, driving in another stake at every chain-length, at the correct distance on each side of the centre stake. A grip about 4 or 5 in. wide should then be cut from stake to stake, so as to mark both the centre and sides of the road upon the ground by a continuous line. The side lines thus set out, it must be remembered, are not the foot of the slopes, but they include 6 ft. on each side for a bank and a ditch. Another stake should therefore be driven at every chain-length, 6 ft. within the outer stakes on each side, and another grip cut to mark the foot of the slopes.

A strong post should next be fixed into the ground,

* The slopes here change from 1 to 1, to 1½ to 1.

upon the centre line, wherever a change in the inclination of the road takes place (as at the 17th stake in the present instance), upon which a cross piece should be placed at the intended height of the formation-surface of the road, and intermediate heights should be put up at such distances as will enable the workmen to keep the embankments to their proper level. For cuttings, pits must be sunk correspondingly, at certain intervals, to the depth of the formation-surface, to serve as guides to the excavators in forming the cutting.

In the foregoing example, the slopes have been taken at ratios of 1 to 1, and $1\frac{1}{2}$ to 1; but it should be remembered that the inclination of the side slopes demands peculiar attention. The proper inclination depends on the nature of the soil, and the action of the atmosphere and of internal moisture upon it. In common soils, as ordinary garden-earth formed of a mixture of clay and sand, compact clay, and compact stony soils, although the side slopes would withstand very well the effects of the weather with a steeper inclination, it is best, according to Professor Mahan, to give them two base to one perpendicular; as the surface of the roadway will, by this arrangement, be well exposed to the action of the sun and air, which will cause a rapid evaporation of the moisture on the surface. Pure sand and gravel may require a greater slope, according to circumstances. In all cases where the depth of the excavation is great, the base of the slope should be increased. It is not usual to use any artificial means to protect the surface of the side slopes from the action of the weather; but it is a precaution which, in the end, will save much labour and expense in keeping the roadway in good order. The simplest means which can be used for this purpose, consist in covering the slopes with good sods, or else with a layer of vegetable mould about 4 inches thick, carefully laid and sown with grass seed. These means are amply sufficient to protect the side slopes from injury when they are not exposed to any other causes of deterioration than the wash of the rain, and the action of frost on the ordinary moisture retained by the soil.

The side slopes form usually an unbroken surface from the foot to the top. But in deep excavations, and particularly in soils liable to slips, they are sometimes formed with horizontal offsets, termed *benches*, which are made a few feet wide, and have a ditch on the inner side to receive the surface-water from the portion of the side slope above them. These benches catch and retain the earth that may fall from the portion of the side slope above.

When the side slopes are not protected, it will be well, in localities where stone is plenty, to raise a small wall of dry stone at the foot of the slopes, to prevent the wash of the slopes from being carried into the roadway.

A covering of brush-wood, or a thatch of straw, may also be used with good effect; but, from their perishable nature, they will require frequent renewal and repairs.

In excavations through solid rock, which does not disintegrate on exposure to the atmosphere, the side slopes might be made perpendicular; but as this would exclude, in a great degree, the action of the sun and air, which is essential to keeping the road-surface dry and in good order, it is necessary to make the side slopes with an inclination, varying from one base to one perpendicular, to one base to two perpendicular, or even greater, according to the locality:—the inclination of the slope on the south side in northern latitudes being the greater, to expose better the road-surface to the sun's rays.

The slaty rocks generally decompose rapidly on the surface, when exposed to moisture and the action of frost. The side slopes in rocks of this character may be cut into steps, and then be covered by a layer of vegetable mould sown in grass seed, or else the earth may be sodded in the usual way.

The stratified soils and rocks, in which the strata have a *dip*, or inclination to the horizon, are liable to *slips*, or to give way, by one stratum becoming detached and sliding on another; which is caused either from the action of frost, or from the pressure of water, which insinuates itself between the strata. The worst soils of this character are those formed of alternate strata of clay and sand; particularly if the clay is of a nature to become semi-fluid when mixed with water. The best preventives that can be resorted to in these cases are, to adopt a system of thorough drainage, to prevent the surface-water of the ground from running down the side slopes, and to cut off all springs which run towards the roadway from the side slopes. The surface-water may be cut off by means of a single ditch made on the up-hill side of the road, to catch the water before it reaches the slope of the excavation, and convey it off to the most convenient natural water-courses; for, in almost every case, it will be found that the side slope on the down-hill side is, comparatively speaking, but slightly affected by the surface-water.

Where slips occur from the action of springs, it frequently become a very difficult task to secure the side slopes. If the sources can be easily reached by excavating into the side slopes, drains formed of layers of fascines, or brush-wood, may be placed to give an outlet to the water, and prevent its action upon the side slopes. The fascines may be covered on top with good sods laid with the grass side beneath, and the excavation made to place the drain be filled in with good earth well rammed. Drains formed of broken stone, covered in like manner on top with a layer of sod to prevent the drain from becoming choked with earth, may be used under the same circumstances as fascine drains. Where the sources are not isolated, and the whole mass of the soil forming the side slopes appears saturated, the drainage may be effected by excavating trenches a few feet wide at intervals to the depth of some feet into the side slopes, and filling them with broken stone, or else a general drain of broken stone may be made throughout the whole extent of the side slope by excavating into it. When this is deemed necessary, it will be well to arrange the drain like an inclined retaining-wall, with buttresses at intervals projecting into the earth further than the general mass of the drain. The front face of the drain should, in this case, also be covered with a layer of sods with the grass side beneath, and upon this a layer of good earth should be compactly laid to form the face of the side slopes. The drain need only be carried high enough above the foot of the side slope to tap all the sources; and it should be sunk sufficiently below the roadway surface to give it a secure footing.

The drainage has been effected, in some cases, by sinking wells or *shafts* at some distance behind the side slopes, from the top surface to the level of the bottom of the excavation, and leading the water which collects in them, by pipes, into drains at the foot of the side slopes. In others, a narrow trench has been excavated, parallel to the axis of the road, from the top surface to a sufficient depth to tap all the sources which flow towards the side slope, and a drain formed either by filling the trench wholly with broken stone, or else by arranging an open conduit at the bottom to receive the water collected, over which a layer of brush-wood is laid, the remainder of the trench being filled with broken stone.

In some instances, the side slopes of very bad soils have been secured by a facing of brick arranged in a manner very similar to the method resorted to for securing the perpendicular sides of narrow deep trenches by a timber-facing. The plan pursued is, to place, at intervals along the excavation, strong buttresses of brick on each side, opposite to each other, and to connect them at bottom by a reversed arch. Between these buttresses are placed, at suitable heights, one or more brick beams, formed at bottom with a flat segment arch, and at top with a like arch inverted. The buttresses, secured in this way, serve as piers for vertical cylindrical arches, which form the facing and support the pressure of the earth between the buttresses.

In forming the embankments, the side slopes should be made with a less inclination than that which the earth naturally assumes; for the purpose of giving them greater durability, and to prevent the width of the top surface, along which the roadway is made, from diminishing by every change in the side slopes, as it would were they made with the natural slope. To protect the side slopes more effectually, they should be sodded, or sown in grass seed; and the surface-water of the top should not be allowed to run down them, as it would soon wash them into gullies, and destroy the embankment. In localities where stone is plentiful, a sustaining wall of dry stone may be advantageously substituted for the side slopes.

To prevent, as far as possible, the settling which takes place in embankments, they should be formed with great care; the earth being laid in successive layers of about four feet in thickness, and each layer well settled with rammers. As this method is very expensive, it is seldom resorted to except in works which require great care, and are of trifling extent. For extensive works, the method usually followed, on account of economy, is to embank out from one end, carrying forward the work on a level with the top surface. In this case, as there must be a want of compactness in the mass, it would be best to form the outsides of the embankment first, and to gradually fill in towards the centre, in order that the earth may arrange itself in layers with a dip from the sides inwards; this will in a great measure counteract any tendency to slips outward. The foot of the slopes should be secured by buttressing them either by a low stone wall, or by forming a slight excavation for the same purpose.

In some cases surface drains, termed *catch-water drains*, are made on the side slopes of cuttings. They are run up obliquely along the surface, and empty directly into the cross drains which convey the water into the natural water-courses.

When the roadway is in side-forming, cross drains of the ordinary form of culverts are made, to convey the water from the side channels and the covered drains into the natural water-courses. They should be of sufficient dimensions to convey off a large volume of water, and to admit a man to pass through them, so that they may be readily cleared out, or even repaired, without breaking up the roadway over them.

The only drains required for embankments are the ordinary side channels of the roadway, with occasional culverts to convey the water from them into the natural water-courses. Great care should be taken to prevent the surface-water from running down the side slopes, as they would soon be washed into gullies by it.

When the axis of the roadway is laid out on the side slope of a hill, and the road-surface is formed partly by excavating and partly by embanking out, the usual and most simple method is to extend out the embankment gradually along the whole line of excavation. This method is insecure, and no pains therefore should be spared to give the embankment a good footing on the natural surface upon which it rests, particularly at the foot of the slope. For this purpose the natural surface should be cut into steps, or offsets, and the foot of the slope be secured by

SIDE-FORMINGS.

buttressing it against a low stone wall, or a small terrace of carefully rammed earth.

In side-formings along a natural surface of great inclination, the method of construction just explained will not be sufficiently secure; sustaining-walls must be substituted for the side slopes, both of the excavations and embankments. These walls may be made simply of dry stone, when the stone can be procured in blocks of sufficient size to render this kind of construction of sufficient stability to resist the pressure of the earth. But when the blocks of stone do not offer this security, they must be laid in mortar, and hydraulic mortar is the only kind which will form a safe construction. The wall which supplies the slope of the excavation should be carried up as high as the natural surface of the ground; the one that sustains the embankment should be built up to the surface of the roadway; and a parapet-wall should be raised upon it, to secure vehicles from accidents in deviating from the line of the roadway.

A road may be constructed partly in excavation and partly in embankment along a rocky ledge, by blasting the rock, when the inclination of the natural surface is not greater than one perpendicular to two base; but with a greater inclination than this, the whole should be in excavation.

There are examples of road constructions, in localities like the last, supported on a frame-work, consisting of horizontal pieces, which are firmly fixed at one end by being let into holes drilled in the rock, and are sustained at the other by an inclined strut underneath, which rests against the rock in a shoulder formed to receive it.

When the excavations do not furnish sufficient earth for the embankments, it is obtained from excavations termed *side-cuttings*, made at some place in the vicinity of the embankment, from which the earth can be obtained with the most economy.

If the excavations furnish more earth than is required for the embankment, it is deposited in what is termed a *spoil-bank*, on the side of the excavation. The spoil-bank should be made at some distance back from the side slope of the excavation, and on the down-hill side of the top-surface; and suitable drains should be arranged to carry off any water that might collect near it and affect the side slope of the excavation.

The forms to be given to side-cuttings and spoil-banks will depend, in a great degree, upon the locality; they should, as far as practicable, be such that the cost of removal of the earth shall be the least possible.

CHAPTER III.

RESISTANCE TO TRACTION ON COMMON ROADS.

The following are the general results of the experiments made by M. Morin upon the resistance to the traction of vehicles on common roads:—

1st. The resistance to traction is directly proportional to the load, and inversely proportional to the diameter of the wheel.

2nd. Upon a paved or a hard macadamized road the resistance is independent of the width of the tire, when this quantity exceeds from 3 to 4 inches.

3rd. At a walking pace, the resistance to traction is the same, under the same circumstances, for carriages with springs and for carriages without springs.

4th. Upon hard macadamized roads and upon paved roads, the resistance to traction increases with the velocity: the increments of traction being directly proportional to the increments of velocity above the velocity 3·28 feet per second, or about 2¼ miles per hour. The equal increments of traction thus due to equal increments of velocity, are less as the road is smoother, and as the carriage is less rigid or better hung.

5th. Upon soft roads, of earth, or sand or turf, or roads fresh and thickly gravelled, the resistance to traction is independent of the velocity.

6th. Upon a well-made and compact pavement of hewn stones, the resistance to traction at a walking pace is not more than three-fourths of the resistance upon the best macadamized roads, under similar circumstances. At a trotting pace, the resistances are equal.

7th. The destruction of the road is, in all cases, greater as the diameters of the wheels are less, and it is greater in carriages without than with springs.

The next experiments which may be quoted, are those of Sir John Macneil,* made with an instrument invented by him for the purpose of measuring the tractive force required on different descriptions of road, to draw a wagon weighing 21 cwt., at a very low velocity. The general results which he obtained are given in the following table:—

TABLE No. 3.—RESULTS OF TRACTION FORCE TO DRAW 21 CWT. ON A LEVEL.
(Sir John Macneil.)

Description of road.	Total tractive force.	Tractive force per ton.
	lbs.	lbs.
On a well-made pavement	33	31·2
On a road made with six inches of broken stone of great hardness, laid either on a foundation of large stones, set in the form of a pavement, or upon a bottoming of concrete	46	44
On an old flint road, or a road made with a thick coating of broken stone, laid on earth	65	62
On a road made with a thick coating of gravel, laid on earth	147	140

Sir John Macneil has also given the following arbitrary formulæ,† for calculating the resistance to traction on level roads of various kinds. They have been deduced from a considerable number of experiments made on the different kinds of road specified below, with carriages moving at various velocities. Putting R for the force required to move the carriage, w the weight of the carriage, w that of the load, all expressed in pounds, v the velocity in feet per second, and c a constant number, which depends upon the

* Sir H. Parnell on Roads, p. 73.   † Ibid., p. 464.

nature of the surface over which the carriage is drawn, and the value of which for several different kinds of road is as follows:—

On a timber surface . . . . . $c =$ 2
On a paved road . . . . . . „ 2
On a well-made broken stone road, in a dry clean state . „ 5
On a well-made broken stone road, covered with dust . „ 8
On a well-made broken stone road, wet and muddy . „ 10
On a gravel or flint road, in a dry clean state . . „ 13
On a gravel or flint road, in a wet and muddy state . „ 32

$$\text{Stage wagon, } R = \frac{W+w}{93} + \frac{w}{40} + cv \quad . \quad . \quad . \quad (1.)$$

$$\text{Stage coach, } R = \frac{W+w}{100} + \frac{w}{40} + cv \quad . \quad . \quad . \quad (2.)$$

RULE 1.—Divide the gross weight of the carriage when loaded, in pounds, by 93 if a wagon, or by 100 if a coach, and to the quotient add one-fortieth of the weight of the load only; to the sum, add the product of the velocity in feet per second, by the proper constant for the particular kind of road. The sum is the force in pounds required to draw the carriage at the given velocity upon that description of road.

For example: What force would be requisite to move a stage-coach weighing 2,060 lbs., and having a load of 1,100 lbs., at a velocity of 9 ft. per second, along a broken-stone road covered with dust? By the rule,

$$\frac{2060 + 1100}{100} + \frac{1100}{40} + (8 \times 9) = 131 \cdot 1 \text{ lbs.}$$

the force required.

To consider, next, the additional resistance which is occasioned when the road, instead of being level, is inclined against the load, in a greater or less degree. In order to simplify the question, suppose the whole weight to be supported on one pair of wheels, and that the tractive force is applied in a direction parallel to the surface of the road. Let A B, Fig. 21, represent a portion of an inclined road, c being a carriage just sustained in its position by a force acting in the direction C D. The carriage is kept in position by three forces, namely, by its own weight W, acting in the vertical direction C F, by the force F, applied in the direction C D parallel to the surface of the road, and by the pressure P, which is exerted by the carriage against the surface of the road acting in the direction C E, perpendicular to the surface.

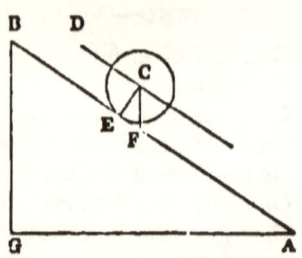

Fig. 21.—Gravity on an inclined plane.

To determine the relative magnitude of these three forces, draw the horizontal line A G, and the vertical line B G; then, since the two lines C F and D G are parallel, and are both cut by the line A B, they must make the two angles C F B and A B G equal; also the two angles C E F and A G B are equal, being both right angles; therefore the remaining angles F C E and B A G are equal, and the two triangles C F E and A B G are similar. And as the three sides of the triangle C F E are proportional to the three forces by which the carriage is sustained, so also are the three sides of the triangle A B G; that is to say, A B, or the length of the road is proportional to W, or the weight of the carriage; B G, or the vertical rise is proportional to F, or the force required to sustain the carriage on the incline; and A G, or the horizontal distance for the rise is proportional to P, or the force with which the carriage presses upon the surface of the road.

Therefore,

$$W : AB :: F : GB,$$
$$\text{and } W : AB :: P : AG,$$

And if A G be made of such a length that the vertical rise, B G, of the road, is exactly one foot, then,—

$$F = \frac{W}{AB} = \frac{W}{\sqrt{AG^2 + 1}} = W \cdot \sin \beta \quad . \quad . \quad . \quad (3.)$$

and $P = \dfrac{W \cdot AG}{AB} = \dfrac{W \cdot AG}{\sqrt{AG^2 + 1}} = w \cdot \cos \beta$ . . . (4.)

in which β is the angle B A G.

These formulæ reduced to verbal rules are as follows:—

RULE 2.—*To find the force requisite to sustain a carriage upon an inclined road (the effects of friction being neglected)*, divide the weight of the carriage, including its load, by the *inclined* length of the road, the vertical rise of which is one foot, and the quotient is the force required.

RULE 2.—*To find the pressure of a carriage against the surface of an inclined road*, multiply the weight of the loaded carriage by the *horizontal* length of the road, and divide the product by the *inclined* length of the same; the quotient is the pressure required.

Example.—What is the force required to sustain a carriage weighing 3,270 lbs. upon a road, the inclination of which is one in thirty, and what is the pressure of the carriage upon the surface of the road?

Here the horizontal length of the road, A G, being equal to 30, for a rise of 1 foot, the inclined length, A B = $\sqrt{AG^2 + 1}$ = 30·017, and by the first rule, 3,270 ÷ 30·017 = 108·93 lbs. for the force required to sustain the carriage on the road.

By the second rule, 3,270 × 30 ÷ 30·017 = 3,269·9 lbs., the pressure of the carriage upon the surface of the road.

Since the pressure of a carriage on a sloping road is found by multiplying its weight by the horizontal length of the road and dividing by the inclined length, and as the former is always less than the latter, it follows that the force with which a carriage bears upon an inclined road is less than its actual weight. In the foregoing example, it is about two pounds less; but, unless the inclination is very steep, it is not necessary to distinguish the difference of pressure, as the pressure may be assumed to be equal to the weight of the carriage.

If the resistance which is to be overcome in moving a carriage, at a given rate, upon a horizontal road, be expressed by R, then R + F is the resistance in ascending a hill, and R − F descending a hill, with the same velocity; neglecting the decrease in the weight of the carriage produced by the inclination of the road. Taking, however, this decrease into consideration, the following modification in the formulæ (1.) and (2.) will be requisite to adapt them to an inclined road:—

$$R = \left(\frac{W+w}{93}+\frac{w}{40}\right) \cdot \cos \beta \mp (W+w) \cdot \sin \beta + c\,v \cdot (5.)$$

in the case of a common stage wagon; and in that of a stage coach,

$$R = \left(\frac{W+w}{100}+\frac{w}{40}\right) \cdot \cos \beta \mp (W+w) \cdot \sin \beta + c\,v \cdot (6.),$$

the upper sign being taken when the vehicle is drawn down the incline, and the lower when it is drawn up the same.

To ascertain the resistance in passing up or down a hill, therefore, the resistance on a level road is first to be calculated, by Rule 1, page 53. To this is to be added the force necessary to sustain the carriage on the incline, in ascending, calculated by Rule 2, page 55; or, in descending, the same force is to be subtracted from the resistance on a level.

As an example, take, as before, the case of a stage coach weighing 2,060 lbs., besides a load of 1,100 lbs., at a velocity of 9 ft. per second, up a broken stone road of which the surface is covered with dust, and which is inclined at the rate of one in thirty.

The force to sustain the coach on this slope is, by Rule 2,

$$\frac{3160}{30} = 105 \cdot 3 \text{ lbs.}$$

Adding this force to the force already found at page 53, requisite to move the same coach on a level road, the sum is (105·3 + 131·1 =) 236·4 lbs., for the force required to move the coach with a velocity of 9 ft. per second *up* the inclined road of one in thirty. To draw the coach *down* the same incline, at the same velocity, the resulting force required is the difference of the two forces already found, or it is (131·1 − 105·3=) 25·8 lb,

The same example worked by formula (6) will give

$$\left(\frac{2060 + 1100}{100}\right) \cdot 9995 + (2060 + 1100) \cdot 0333 + (8 \times 9)$$

= 236·3 lbs, when the carriage is drawn up the incline; and

$$\left(\frac{2060 + 1100}{100}\right) \cdot 9995 - (2060 + 1100) \cdot 0333 + (8 \times 9)$$

= 25·84 lbs., when the carriage is drawn down the incline, the result being the same as that given by the rule.

The following table has been calculated in order to show, with sufficient exactness for most practical purposes, the force required to draw carriages over inclined roads, and the comparative advantage of such roads and those which are perfectly level. The first column expresses the rate of inclination, and the second the equivalent angle; the two next columns contain the force requisite to draw a common stage wagon weighing with its load 6 tons, at a velocity of 4·4 ft. per second (or 3 miles per hour) along a macadamized road in its usual state, both when ascending and descending the hill; the fifth and sixth columns contain the length of level road which would be equivalent to a mile in length of the inclined road, that is, the length of level road which would require the same mechanical work to be expended in drawing the wagon over it, as would be necessary to draw the wagon over a mile of the inclined road. The next four columns contain the same information as the four just described, with reference to a stage coach supposed to weigh with its load 3 tons, and to travel at the rate of 8·8 ft. per second, or 6 miles per hour.

TABLE No. 4.—RESISTANCE TO TRACTION ON INCLINED ROADS.

Rate of Inclination.	Angle with the Horizon.	For a Stage Wagon, 6 tons gross.				For a Stage Coach, 3 tons gross.			
		Force required to draw the wagon up the incline.	Force required to draw the wagon down the incline.	Equivalent length of level road for an ascending wagon.	Equivalent length of level road for a descending wagon.	Force required to draw the coach up the incline.	Force required to draw the coach down the incline.	Equivalent length of level road for an ascending coach.	Equivalent length of level road for a descending coach.
	° ′ ″	lbs.	lbs.	Miles.	Miles.	lbs.	lbs.	Miles.	Miles.
1 in 600	0 5 44	286	241	1·085	·9150	373	350	1·030	·9690
,, 575	0 5 59	287	240	1·088	·9116	373	350	1·032	·9676
,, 550	0 6 15	288	239	1·093	·9074	374	349	1·033	·9662
,, 525	0 6 33	289	238	1·097	·9029	374	349	1·035	·9646
,, 500	0 6 53	291	237	1·102	·8979	375	348	1·037	·9629
,, 475	0 7 14	292	235	1·107	·8926	376	347	1·039	·9605
,, 450	0 7 38	294	334	1·113	·8869	377	347	1·041	·9588
,, 425	0 8 5	295	232	1·120	·8801	377	346	1·043	·9563
,, 400	0 8 36	297	230	1·128	·8725	378	345	1·046	·9535
,, 375	0 9 10	300	228	1·136	·8642	380	344	1·049	·9505
,, 350	0 9 49	302	225	1·146	·8543	381	342	1·053	·9469
,, 325	0 10 35	305	222	1·157	·8433	382	341	1·056	·9430
,, 300	0 11 28	309	219	1·170	·8301	384	339	1·061	·9381
,, 290	0 11 51	310	217	1·176	·8245	385	338	1·064	·9358
,, 280	0 12 17	312	216	1·182	·8179	386	338	1·066	·9336
,, 270	0 12 44	314	214	1·189	·8111	386	337	1·068	·9314
,, 260	0 13 13	315	212	1·196	·8039	387	336	1·071	·9286
,, 250	0 13 45	317	210	1·204	·7963	388	335	1·074	·9259
,, 240	0 14 19	320	208	1·212	·7876	390	334	1·077	·9226
,, 230	0 14 57	322	205	1·222	·7785	391	332	1·080	·9192
,, 220	0 15 37	325	203	1·232	·7683	392	331	1·084	·9156
,, 210	0 16 22	328	200	1·243	·7573	394	330	1·088	·9115
,, 200	0 17 11	331	197	1·255	·7451	395	328	1·092	·9071
,, 190	0 18 6	334	193	1·268	·7319	397	326	1·097	·9024
,, 180	0 19 6	338	189	1·283	·7171	399	324	1·103	·8968
,, 170	0 20 13	343	185	1·300	·7004	401	322	1·109	·8908
,, 160	0 21 29	848	180	1·319	·6814	404	320	1·116	·8839
,, 150	0 22 55	353	174	1·341	·6587	406	317	1·123	·8761
,, 140	0 24 33	360	168	1·364	·6359	410	314	1·132	·8673
,, 130	0 26 27	367	160	1·392	·6079	413	310	1·142	·8573
,, 120	0 28 39	376	152	1·425	·5752	418	306	1·154	·8451
,, 110	0 31 15	386	142	1·451	·5491	423	300	1·169	·8308
,, 100	0 34 23	398	129	1·510	·4903	429	294	1·185	·8142
,, 95	0 86 11	405	122	1·537	·4634	432	291	1·195	·8045
,, 90	0 38 12	413	114	1·566	·4338	436	287	1·206	·7937
,, 85	0 40 27	422	106	1·600	·4004	441	282	1·219	·7801
,, 80	0 42 58	432	96	1·637	·3629	446	278	1·232	·7677

TABLE.

Table No. 4.—(Continued.)

Rate of Inclination.	Angle with the Horizon.			For a Stage Wagon. 6 tons gross.				For a Stage Coach. 3 tons gross.			
	°	′	″	Force required to draw the wagon up the incline.	Force required to draw the wagon down the incline.	Equivalent length of level road for an ascending wagon.	Equivalent length of level road for a descending wagon.	Force required to draw the coach up the incline.	Force required to draw the coach down the incline.	Equivalent length of level road for an ascending coach.	Equivalent length of level road for a descending coach.
				lbs.	lbs.	Miles.	Miles.	lbs.	lb s.	Miles.	Miles.
1 in 75	0	45	51	443	85	1·680	·3204	451	272	1·217	·7522
,, 70	0	49	7	456	72	1·728	·2719	457	265	1·265	·7345
,, 65	0	52	54	470	57	1·784	·2161	465	258	1·285	·7113
,, 60	0	57	18	488	40	1·850	·1505	474	250	1·309	·6903
,, 55	1	2	30	508	19	1·926	·0736	484	239	1·337	·6620
,, 50	1	8	6	533	—	2·019	—	496	227	1·371	·6283
,, 45	1	16	24	562	—	2·133	—	511	212	1·412	·5871
,, 40	1	25	57	600	—	2·274	—	530	194	1·464	·5354
,, 35	1	38	14	648	—	2·456	—	554	170	1·530	·4690
,, 34	1	41	8	659	—	2·499	—	559	164	1·546	·4535
,, 33	1	44	12	671	—	2·544	—	565	158	1·562	·4370
,, 32	1	47	27	684	—	2·593	—	572	152	1·580	·4193
,, 31	1	50	55	697	—	2·644	—	578	145	1·599	·4007
,, 30	1	54	37	712	—	2·699	—	586	138	1·619	·3805
,, 29	1	58	34	727	—	2·758	—	593	130	1·640	·3592
,, 28	2	2	5	744	—	2·820	—	602	122	1·663	·3363
,, 27	2	7	2	762	—	2·888	—	610	113	1·688	·3119
,, 26	2	12	2	781	—	2·960	—	620	103	1·714	·2854
,, 25	2	17	26	801	—	3·038	—	630	93	1·743	·2566
,, 24	2	23	10	823	—	3·120	—	641	82	1·774	·2257
,, 23	2	29	22	847	—	3·213	—	653	69	1·808	·1919
,, 22	2	36	10	874	—	3·313	—	666	56	1·844	·1554
,, 21	2	43	35	903	—	3·423	—	681	42	1·884	·1150
,, 20	2	51	21	933	—	3·538	—	696	26	1·926	·0730
,, 19	3	0	46	970	—	3·677	—	714	8	1·977	·0221
,, 18	3	10	47	1009	—	3·826	—	734	—	2·032	—
,, 17	3	21	59	1053	—	3·991	—	756	—	2·092	—
,, 16	3	34	35	1102	—	4·178	—	780	—	2·160	—
,, 15	3	48	51	1157	—	4·388	—	807	—	2·234	—
,, 14	4	5	14	1221	—	4·629	—	839	—	2·322	—
,, 13	4	23	56	1294	—	4·906	—	875	—	2·423	—
,, 12	4	45	49	1379	—	5·229	—	918	—	2·540	—
,, 11	5	11	40	1480	—	5·611	—	968	—	2·679	—
,, 10	5	42	58	1600	—	6·067	—	1028	—	2·846	—
,, 9	6	20	25	1747	—	6·623	—	1101	—	3·048	—
,, 8	7	7	30	1929	—	7·315	—	1192	—	3·300	—
,, 7	8	7	48	2162	—	8·199	—	1308	—	3·621	—

The foregoing table may be considered as affording a view of the comparative disadvantage of hilly roads with light and heavy traffic; the stage wagon weighing 6 tons and travelling at the speed of 3 miles per hour, may be taken as a fair average for goods traffic, and the stage coach, weighing 3 tons and running 6 miles an hour, for passenger traffic. It is shown that the resistance on hills is much more unfavourable to the wagon than to the coach. The force which would be requisite to move the wagon on a level road would be 264 lbs., and that to move the coach 362 lbs., being an excess of 98 lbs. for the traction of the coach. But, with a road inclined at the rate of 1 in 600, this excess is only (373 − 286 =) 87 lbs.; and when the inclination of the road amounts to about 1 in 70, the forces required to draw them become equal. As the inclination of the road increases beyond this, the excess of force requisite to draw the waggon over that necessary to move the coach, increases rapidly until, at an inclination of 1 in 7, it amounts to (2162 − 1308 =) 854 lbs.

Comparing the forces required to draw either the wagon or the coach up and down any given incline, the former is as much greater than the force required on a level road as the latter is less. It might thence be concluded that, when a vehicle passes alternately each way along the road, no real loss is occasioned by the inclination of the road, since as much power is gained in the descent of the hill as is lost in its ascent. Such is not, however, practically the fact, for whilst it is necessary in the ascending journey to have either a greater number of horses, or more powerful horses, than would be requisite if the road were entirely level, no corresponding reduction can be made in the descending journey. There must be horses sufficient to draw the vehicle along the level portions of the road; nor, generally speaking, have the horses less to do in descending the hill,

ANGLE OF REPOSE.

since they frequently are required to push back, to prevent the speed of the coach from being accelerated to a rate beyond the limits of safety.

In a practical sense, therefore, it may be considered that the fifth and ninth columns in the foregoing table express the length of level road which would be equivalent to a mile of road with the stated inclination, the fifth giving the result for heavy traffic, and the ninth for passenger traffic. For instance, against the incline 1 in 75, there is a length of 1·247 miles, or about a mile and a quarter, in the ninth column, given as the equivalent length of level road for 1 mile of ascent on the incline, in the sense that the same quantity of work of traction would be requisite to move a coach of 3 tons, at a velocity of 6 miles per hour, along one as along the other. But, in other respects, the incline might be more advantageous than the level; for instance, the shorter road would cost less for repair, and would be passed over in less time. The table, therefore, merely expresses the equivalent length as far as the mechanical work required for the traction is concerned.

From the results of Sir John Macneil's experiments on tractional resistance, page 52, Professor Mahan deduces that the angle of repose in the first case is represented by $\frac{33}{2240}$, or 1 in 68 nearly; and that the slope of the road should therefore not be greater than one perpendicular to sixty-eight in length; or that the height to be ascended must not be greater than one sixty-eighth of the distance between the two points measured along the road, in order that the force of friction may counteract that of gravity in the descent of the road.

A similar calculation will show that the angle of repose in the other cases will be as follows:—

   No. 2, . . . 1 to 35 nearly.
   ,, 3, . . . 1 to 15 ,,
   ,, 4 and 5, . . 1 to 49 ,,

These numbers, which give the angle of repose between 1 in 35 and 1 in 49 for the kinds of road-covering, Nos. 2 and 4, in most ordinary use, and corresponding to a road-surface in good order, may be somewhat increased, to from 1 in 28 to 1 in 33, for the ordinary state of the surface of a well-kept road, without there being any necessity for applying a brake to the wheels in descending, or going out of a trot in ascending. The steepest gradient that can be allowed on roads with a broken-stone covering is about 1 in 20, as this, from experience, is found to be about the angle of repose upon roads of this character in the state in which they are usually kept. Upon a road with this inclination, a horse can draw, at a walk, his usual load for a level without requiring the assistance of an extra horse; and experience has further shown that a horse at the usual walking pace will attain, with less apparent fatigue, the summit of a gradient of 1 in 20 in nearly the same time that he would require to reach the same point on a trot over a gradient of 1 in 33.

A road on a dead level, or one with a continued and uniform ascent between the points of arrival and departure, where they lie upon different levels, is not the most favourable to the draft of the horse. Each of these seems to fatigue him more than a line of alternate ascents and descents of slight gradients; as, for example, gradients of 1 in 100, upon which a horse will draw as heavy a load with the same speed as upon a horizontal road.

The gradients should in all cases be reduced as far as practicable, as the extra exertion that a horse must put forth in overcoming heavy gradients is very considerable; they should, as a general rule, therefore, be kept as low at least as 1 in 33, wherever the ground will admit of it. This can generally be effected, even in ascending steep hill-sides, by giving the axis of the road a zig-zag direction, connecting the straight portions of the zig-zags by circular arcs. The gradients of the curved portions of the zig-zags should be reduced, and the roadway also at these points should be widened, for the safety of vehicles descending rapidly. The width of the road may be increased about one-fourth, when the angle between the straight portions of the ziz-zags is from 120° to 90°; and the increase should be nearly one-half where the angle is from 90° to 60°.

NOTE BY THE EDITOR.—Sir John Macneil, in 1836, maintained that no road was perfect unless its gradients were equal to or less than 1 in 40. In thus limiting the ruling gradient to 1 in 40, he justifies the assertion by the much greater outlay for repair on roads of steeper gradients. For instance, he adduces as a fact not generally known, that if a road has no greater inclinations than 1 in 40, there is 20 per cent. less cost for maintenance than for a road having an inclination of 1 in 20. The additional cost is due not only to the greater injury by the action of horses' feet on the steeper incline, which has already been noticed, but also to the greater fatigue of the road by the more frequent necessity for sledging or braking the wheels of vehicles in descending the steeper portions.

Professor Mahan, it has been seen, page 62, recommends, as a general rule, that the gradients should be kept as low as 1 in 33; whilst M. Dumas, engineer-in-chief of the French Ponts et Chaussées, writing in 1843,[*] recommended, as a maximum rate of inclination, 1 in 50; for, he says, "not only are the surfaces of steeply-inclined roads subjected to abrasion by the feet of horses clambering up the hill, but, in the intervals of rest, loose stones are placed as props behind the wheels of vehicles, which are usually allowed to remain where they have been temporarily placed, and may be the causes of serious accidents."

[*] "Annales des Ponts et Chaussées," 2nd series, 1 Semestre, 1843, page 343.

Besides, he states as the result of experience, that on broken-stone roads, in perfect condition, the resistance to traction is 1-50th of the gross weight, or 45 lbs. per ton, for which the angle of repose is 1 in 50; and he adds, with scientific acuteness, "that for the ascent of an incline of 1 in 50, the traction force required is just double that which is required on the level." "Evidently," he continues, "there is no danger, under such conditions, in making the descent, since it requires but the slightest effort to check the vehicle; whilst, in ascending, the horses can, without trouble, exert double the customary force for a short time." In fact, horses can easily enough surmount gradients of more than 3 per cent., or 1 in 33, at a trot, on roads in mediocre condition.

M. Dupuit recommends for the maximum gradients of roads—

    For metalled roads . . 3 per cent. or 1 in 33
    For pavements . . . 2 ,,   1 in 50

It can but be observed, upon the foregoing evidence, that Sir John Macneil's proportion of 1 in 40 for the maximum slopes of roads, is most nearly an average of the deductions which have been cited.

But there is another condition—the minimum longitudinal slope of a road. It should not be quite level, for provision must be made, by inclining the road, for running off surface-water. The minimum slope is fixed by one authority at 1 in 80; by another, at half a degree, or 1 in 115; and by the Corps des Ponts et Chaussées, at 1 in 125.

In the Second Part of this work, by the Editor, he has given an analysis of the Rolling or Circumferential Resistance of Wheels.

CHAPTER IV.

ON THE SECTION OF ROADS.

WHERE hills or gradients are unavoidable, they should be made as easy as possible; and, although a certain amount of additional power must be required to draw a carriage up a hill, compared with the resistance on a level, yet so long as the inclination is within a certain limit, the hilly road may be considered as safe as a level road. This limit depends upon the nature and condition of the surface of the road, and it is attained in any particular case when the inclination of the road is made equal to the limiting angle of resistance for the materials composing its surface; that is, when it is such that a carriage once set in motion on the road, would just continue its descent without any additional force being applied. When this limit is exceeded, the carriage descends with an accelerated velocity, unless the horses or other force be employed to restrain it; and, although, in such a case, the use of a drag, by increasing the resistance, would in a measure obviate the danger, yet the injury done to the surface of the road by the use of the drag renders it desirable to avoid the use of it altogether. The following table, taken from the second volume of the "Rudiments of Civil Engineering," shows the rate of inclination at which this limit is attained on the various kinds of roads mentioned in the first column.

The values of the resistances on which this table is calculated are those given by Sir John Macneil:—

Description of the road.	Force in lbs. required to move a ton.	Limiting angle of resistance.	Greatest inclination which should be given to the road.
Well-laid pavement	35	0° 50′	1 in 68
Broken stone surface on a bottom of rough pavement or concrete	46	1 11	1 in 49
Broken stone surface laid on an old flint road	65	1 40	1 in 34
Gravel road	147	3 45	1 in 15

The following table of gradients is of considerable value in laying out and arranging roads. The first column contains the gradient, expressed in the ratio of the height to the length; the second and third columns contain the vertical rise in a mile and a chain respectively; the fourth column, the angle of inclination with the horizon; and the last column, the sine of the same angle, which is inserted for facilitating the calculation of the resistances occasioned by the gradient.

TABLE No. 5.—GRADIENTS AND ANGLES OF INCLINATION OF ROADS.

Gradient.	Vertical rise in a mile.	Vertical rise in a chain.	Angle (β) which gradient makes with the horizon.	Sine of angle β.	Gradient.	Vertical rise in a mile.	Vertical rise in a chain.	Angle (β) which gradient makes with the horizon.	Sine of angle β.
			° ′ ″					° ′ ″	
1 in 10	528·0	6·60	5 42 58	·09960	1 in 60	88·0	1·10	0 57 18	·01667
,, 11	480·0	6·00	5 11 40	·09054	,, 65	81·2	1·02	0 52 54	·01539
,, 12	440·0	5·50	4 45 59	·08309	,, 70	75·4	·94	0 49 7	·01429
,, 13	406·1	5·08	4 23 56	·07670	,, 75	70·4	·88	0 45 51	·01334
,, 14	377·1	4·71	4 5 14	·07128	,, 80	66·0	·82	0 42 58	·01250
,, 15	352·0	4·40	3 48 51	·06652	,, 85	62·1	·78	0 40 27	·01177
,, 16	330·0	4·12	3 34 35	·06238	,, 90	58·7	·73	0 38 12	·01111
,, 17	310·6	3·88	3 21 59	·05872	,, 95	55·6	·69	0 36 11	·01053
,, 18	293·3	3·67	3 10 47	·05547	,, 100	52·8	·66	0 34 23	·01000
,, 19	277·9	3·47	3 0 46	·05256	,, 110	48·0	·60	0 31 15	·00909
,, 20	264·0	3·30	2 51 21	·04982	,, 120	44·0	·55	0 28 39	·00833
,, 21	251·4	3·14	2 43 35	·04757	,, 130	40·6	·51	0 26 27	·00769
,, 22	240·0	3·00	2 36 10	·04541	,, 140	37·7	·47	0 24 33	·00714
,, 23	229·6	2·87	2 29 22	·04344	,, 150	35·2	·44	0 22 55	·00666
,, 24	220·0	2·75	2 23 10	·04163	,, 160	33·0	·41	0 21 29	·00625
,, 25	211·2	2·64	2 17 26	·03997	,, 170	31·1	·39	0 20 13	·00588
,, 26	203·1	2·54	2 12 2	·03840	,, 180	29·3	·37	0 19 6	·00556
,, 27	195·5	2·42	2 7 2	·03694	,, 190	27·8	·35	0 18 6	·00527
,, 28	188·5	2·36	2 2 5	·03551	,, 200	26·4	·33	0 17 11	·00500
,, 29	182·1	2·28	1 58 34	·03448	,, 210	25·1	·31	0 16 22	·00476
,, 30	176·0	2·20	1 54 37	·03333	,, 220	24·0	·30	0 15 37	·00454
,, 31	170·3	2·13	1 50 55	·03226	,, 230	23·0	·29	0 14 57	·00435
,, 32	165·0	2·06	1 47 27	·03125	,, 240	22·0	·27	0 14 19	·00417
,, 33	160·0	2·00	1 44 12	·03031	,, 250	21·1	·26	0 13 45	·00400
,, 34	155·3	1·94	1 41 8	·02941	,, 260	20·3	·25	0 13 13	·00385
,, 35	150·9	1·88	1 38 14	·02857	,, 270	19·6	·24	0 12 44	·00370
,, 36	146·7	1·86	1 35 28	·02777	,, 280	18·9	·24	0 12 17	·00357
,, 37	142·7	1·78	1 32 53	·02702	,, 290	18·2	·23	0 11 51	·00345
,, 38	138·9	1·74	1 30 27	·02631	,, 300	17·6	·22	0 11 28	·00334
,, 39	135·4	1·69	1 28 8	·02563	,, 325	16·2	·20	0 10 35	·00308
,, 40	132·0	1·65	1 25 57	·02500	,, 350	15·1	·19	0 9 49	·00286
,, 41	128·8	1·61	1 23 50	·02438	,, 375	14·0	·18	0 9 10	·00267
,, 42	125·7	1·57	1 21 50	·02380	,, 400	13·2	·17	0 8 36	·00250
,, 43	122·8	1·53	1 19 56	·02325	,, 425	12·4	·16	0 8 5	·00235
,, 44	120·0	1·50	1 18 7	·02272	,, 450	11·7	·15	0 7 38	·00222
,, 45	117·3	1·47	1 16 24	·02222	,, 475	11·5	·14	0 7 14	·00210
,, 46	114·8	1·44	1 14 43	·02173	,, 500	10·6	·13	0 6 53	·00200
,, 47	112·3	1·40	1 13 8	·02127	,, 525	10·1	·12	0 6 33	·00191
,, 48	110·0	1·37	1 11 37	·02083	,, 550	9·6	·12	0 6 15	·00182
,, 49	107·7	1·35	1 10 9	·02040	,, 575	9·2	·11	0 5 59	·00174
,, 50	105·6	1·32	1 8 6	·01981	,, 600	8·8	·11	0 5 44	·00167
,, 55	96·0	1·20	1 2 30	·01818					

Width and Transverse Section of Roads.—It is recommended that roads should be wide. It is an error to suppose that the cost of repairing a road depends entirely upon the extent of its surface, and increases with its width. The cost per mile of road depends more upon the extent and the nature of the traffic; and it may be asserted, generally, that the same quantity of material is necessary for the repair of a road, whether wide or narrow, subjected to the same amount of traffic. On the narrow road, the traffic, being confined very much to one track, the road would be worn more severely than when the traffic is spread over a larger surface. The expense of spreading the material over the wider road would be somewhat greater, but the cost for material might be taken as the same. One of the advantages of a wide road is, that the air and the sun exercise more influence in keeping its surface dry than that of a narrow road. The first cost of a wide road is certainly greater than that of a narrow road,—nearly in the ratio of the widths.

For roads situated between towns of importance, and exposed to much traffic, the width should not be less than 30 ft., which would admit of four vehicles abreast; besides a footpath of 6 ft. In the immediate vicinity of large towns and cities, the width should be greater.

The form of the cross section of a road is a subject of much importance, and it is one upon which much difference of opinion exists. Some persons advocate a considerable degree of curvature in the upper surface of the road, with the view of facilitating the drainage of its surface; whilst others are averse to a road being much curved. It is the practice of others, again, to form the road on a flat surface transversely; whilst others give a dip to the formation-surface each way from the centre, supposing that the drainage of the road is thereby facilitated.

The only advantage resulting from the curving of the

transverse section of the road is, that the water, which would otherwise collect upon its surface, is allowed to drain freely off into the side ditches. It has been urged that, in laying on fresh material upon a road, it is necessary to keep the centre much higher than the sides; because, in consequence of the greater number of carriages using the middle of the road, that portion wears more quickly than the sides, and that, unless it is made originally much higher, when so worn it necessarily forms a hollow or depression, from which water cannot drain. Now it is entirely overlooked by those who advance this argument, that the cause of carriages using the middle in preference to the sides of a road, is its rounding form, since it is only in that situation that a carriage stands upright. If the road were comparatively flat, every portion would be equally used; but on very convex roads, the middle is the only portion of the road on which it is safe to travel. On this subject, Mr. Macadam remarks, in his evidence before a committee of the House of Commons,* "I consider a road should be as flat as possible with regard to allowing the water to run off it at all, because a carriage ought to stand upright in travelling as much as possible. I have generally made roads 3 in. higher in the centre than I have at the sides, when they are 18 ft. wide; if the road be smooth and well made, the water will run off very easily in such a slope." And, in answer to the question, "Do you consider a road so made will not be likely to wear hollow in the middle, so as to allow the water to stand, after it has been used for some time?" he replies,— "No; when a road is made flat, people will not follow the middle of it as they do when it is made extremely convex. Gentlemen will have observed that in roads very convex, travellers generally follow the track in the middle, which

* Parliamentary Report on the Highways of the Kingdom, 1819, page 22.

is the only place where a carriage can run upright, by which means three furrows are made by the horses and the wheels, and water continually stands there; and I think that more water actually stands upon a very convex road than on one which is reasonably flat." On the same subject, Mr. Walker remarks,* "A road much rounded is dangerous, particularly if the cross section approaches towards the segment of a circle, the slope in that case not being uniform, but increasing rapidly from the nature of the curve, as we depart from the middle or vertical line. The over-rounding of roads is also injurious to them, by either confining the heavy carriages to one track in the crown of the road, or, if they go upon the sides, by the greater wear they produce, from their constant tendency to move down the inclined plane, owing to the angle which the surface of the road and the line of gravity of the load form with each other; and, as this tendency is perpendicular to the line of draught, the labour of the horse and the wear of the carriage wheels are both much increased by it." †

The drainage of the surface of the road is then the only useful purpose answered by making it convex. But the surface of a road is much more efficiently drained by a small inclination in the direction of its length, than by a much greater transverse slope. On this subject, Mr. Walker has very justly remarked, ‡ "Clearing the road of water is best secured by selecting a course for the road which is not horizontally level, so that the surface of the road may, in its longitudinal section, form, in some degree, an inclined plane; and when this cannot be obtained, owing to the extreme flatness of the country, an artificial

---

* Parliamentary Report, 1819, page 49.

† Remarks on the evils of "barreled roads," as they were called, have been made in the Historical chapter, page 4.—EDITOR.

‡ Parliamentary Report, 1819, page 48.

inclination may generally be made. When a road is so formed, every wheel-track that is made, being in the line of inclination, becomes a channel for carrying off the water much more effectually than can be done by a curvature in the cross section or rise in the middle of the road, without the danger or other disadvantages which necessarily attend the rounding a road much in the middle. I consider a fall of about 1½ inches in 10 feet to be a minimum in this case, if it is attainable without a great deal of extra expense." Whilst, then, the advantages attending the extreme convexity of roads is so small, the disadvantages are considerable. On roads so constructed, vehicles must either keep to the crown of the road, and so occasion an excessive and unequal wear of its surface, or use the sides, with the liability of being overturned. The evidence of coachmasters and others, taken before the committee of the House of Commons, and appended to the report already quoted from, fully bears out the view here taken, and shows that many accidents have arisen from the practice of forming roads with an excessive amount of convexity.

With reference to the above remarks, it is only intended to express disapproval of the practice of forming roads with cross sections rounding in an extreme degree, and not to advocate a perfectly, or nearly, flat road, as many, who have fallen into the opposite error, have done. It is recommended, as the best form which could be given to a road, that its cross section should be formed of two straight lines inclined at the rate of about 1 in 30, and connected at the middle or crown of the road by a segment of a circle, having a radius of about 90 feet. This form of section is shown in Fig. 22, and the rate of inclination there given is quite sufficient to keep the surface of a road drained, provided it is maintained in good order, free from ruts. If the maintenance is neglected, no degree of convexity which can be given to the road will be of any avail, as the water will remain in the hollows or furrows.

The form of cross section here suggested is equally adapted to all widths of road, as the straight lines have merely to be extended at the same rate of inclination, until they meet the sides of the road.

Professor Mahan is of the same opinion with respect to the proper section of a road—namely, that it should be formed of two straight sides, connected at the middle by a flat circular arc. The slope which he recommends is 1 in 48, or 1 inch in 4 feet.

With regard to the form which should be given to the bed upon which the road is to be formed, a similar difference of opinion exists as to whether it should be flat or rounding. Except where the surface upon which the road is to be formed is a strong clay, or other soil impervious to water, no benefit results as far as drainage is concerned, in making the formation-surface or bed of the road convex. It should be borne in mind that, after the road materials are laid upon the formation-surface, and have been for some time subjected to the pressure of heavy vehicles passing over them, they become, to a certain extent, intermixed: the road materials are forced down into the soil, and the soil works up amongst the stones, and the original line of separation becomes entirely lost. If the surface upon which the road materials are laid were to remain a distinct flat surface, perfectly even and regular, into which the road materials could not be forced, then it would be useful to give such an inclination to it as would allow any water which might find its way through the crust or covering of the road, to run off to the sides. Even so, it would have to force a passage between the road materials and the surface on which they rest. Such is, however, far from being the case; and, therefore, unless under peculiar circumstances, no water which finds its way through the hard compact surface of the road itself is arrested by the comparatively soft surface of its bed, and carried off into the side ditches, whatever the slope which might be given to the bed. While, however, it is believed, that, as far as drainage is concerned, it is useless to form the bed or formation surface of the road with a transverse slope, it should, nevertheless, be formed to the same outline as that recommended for the outer surface; making the two surfaces parallel, and thus bestowing an equal depth of road material over every portion of the road. Nevertheless, some road-makers not only recommend a less depth of road materials to be put on the sides than on the middle of the road, but they further advise that an inferior description of material should be employed at the sides. On this subject the following remarks of Mr. Hughes are very much to the purpose:*—"A very common opinion is, that the depth of material in the middle of the road should be greater than at the sides, but, for my part, I have never been able to discover why the sides of the road should be at all inferior to the middle in hardness and solidity. On the contrary, it would be a great improvement in general travelling, if carriages could be made to adhere more strictly to the rule of keeping the proper side of the road; and the reasonable inducement to this practice is, obviously, to make the sides equally hard and solid with the middle. In many roads, even where considerable traffic exists, the only good part of the road consists of about 8 or 10 feet in the middle, the sides being formed with small gravel quite unfit to carry heavy traffic; and the consequence is, that the whole crowd of vehicles is forced into the centre track of the road; thus at least doubling or trebling the wear and tear which would take place if

* "The Practice of Making and Repairing Roads," by Thomas Hughes, 1838, page 12.

the sides were, as they ought to be, equally good with the centre. Another mischievous consequence is, that when it becomes necessary to repair the centre of the road, the carriages are driven off the only good part on to the sides, which consist of weak material, and are often even dangerous for the passage of heavily-laden stage coaches. On the other hand, if equal labour and materials be expended on the whole breadth of the road, it is evident that the wear and tear will be far more uniform; and when any one part requires repair, the traffic may with safety be turned on to another part. Hence, I should always lay on the same depth of material all over the road: and this alone will of course render it necessary to curve the bed of the road."

Great attention should be paid to the drainage of roads, with respect to their upper surface as well as to the surface of the ground on which they rest. To promote the surface-drainage, the road should be formed with the transverse section shown in Fig. 22, and on each side of the road a ditch should be formed of sufficient capacity to receive all the water which can fall upon the road, and it should be of such a depth and with such a declivity as to conduct the water freely away. When footpaths are to be constructed on the sides of the road, a channel or watercourse should be formed between the footpaths and the road, and small drains, formed of tiles or earthern tubes, such as are used for underdraining lands, should be laid under the footpath, at such a level as to take off all the water which may collect in this channel, and convey it into the ditch. In the best-constructed roads, these side channels are paved with flints or pebbles. The drains under the footpath should be introduced about every 60 feet, and should have the same inclination—namely, 1 in 30, as is recommended for the sides of the road, as shown in Fig. 22. A greater inclination would be objec-

ON THE SECTION OF ROADS.

tionable. It is a very frequent mistake to give too great a fall to small drains, for such a current through them is produced as may wash away or undermine the ground around them, and ultimately cause their destruction. When a drain is once closed by any obstruction, no amount of fall which could be given to it would suffice again to clear the passage; whilst a drain having a considerable current through it, would be much more likely to be stopped by foreign matter carried into it, than a drain with a less rapid stream.

When the surface of a road, constructed of suitable materials, compactly laid, is drained in the manner which has just been described, very little water finds its way to the substratum. For some descriptions of soil, however, it is desirable to adopt additional means for maintaining the foundation of a road in a dry state; as, for instance, when the surface is a strong clay through which no water can percolate, or when the ground beneath the road is naturally of a soft, wet, or peaty nature. Under such circumstances a species of underdrainage should be provided. When the surface of the ground is formed to the level intended for the reception of the road materials, trenches should be cut across the road from a foot to eighteen inches in depth, and about a foot wide at the bottom, the sides being sloped as shown in Fig. 23. The distances at which these drains should be formed depends in a great measure on the nature of the soil; in the case of a strong clay soil, or a soil which is naturally very wet, there should be a cross

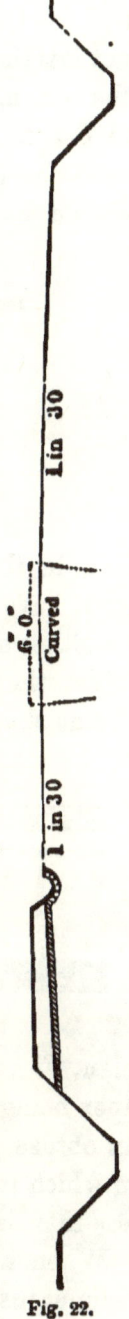

Fig. 22.

drain at intervals of 20 ft. As the ground becomes firmer and drier, the interval may be increased in length. A drain not less than 4 inches square internally should be formed in the trench, of old bricks, drain-tiles, or flat stones, as shown in Fig. 23; or the drain may be formed by any other mode used for under-drains; and the remainder of the trench should be filled with coarse stones free from all clay or

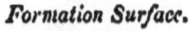

Fig. 23.

dirt, in the manner shown in the figure. These drains must have a fall from the centre of the road into the ditches on either side: an inclination of 1 in 30 will be sufficient. When the road is level in the direction of its length, the drains should run square across; but, on inclined portions of the road, the drains should be formed, as shown on the

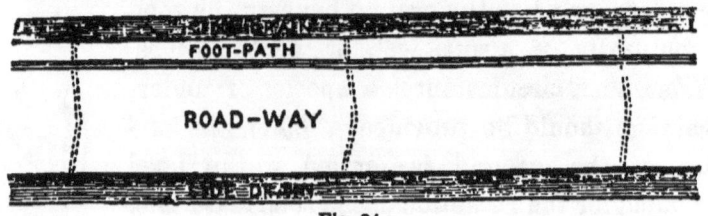

Fig. 24.

plan, Fig. 24, somewhat in the form of a very flat V, the apex being in the middle of the road, and the limbs making an obtuse angle with the line of the road, in the direction in which it falls. The amount of this angle should not be less than is shown in Fig. 24.

When a road with foothpaths is underdrained in the manner just described, it is not necessary to form drains from

ON THE SECTION OF ROADS. 77

the side channel under the footpath into the ditch, as shown in Fig. 22. It is sufficient to carry up a little shaft, constructed in the same way as the drain, from the drain to the channel, covering the shaft with a small grating to prevent leaves or other substances, which might choke the drain, from being carried into it. This method of forming the drains is shown at A in Fig. 25.

The footpath should be not more than 9 inches higher than the bottom of the side channel. The surface of the footpath should have a pitch of 2 inches, towards the side channels, to convey its surface-water into them. When the natural soil is firm and sandy, or gravelly, its surface will serve for the footpath; but in other cases the natural soil must be thrown out to a depth of 6 in., and the excavation filled with fine clean gravel. To prevent the footpath from being damaged by the current of water in the side channel, its side slope, next to the side channel, must be protected by a facing of good sods, or of dry stone.

Independently of the drainage for marshy soils, they will require, when the subsoil is of a spongy elastic nature, an artificial bed for the road-covering. This bed may, in some cases, be formed by simply removing the upper stratum to a depth of several feet, and supplying its place with well-packed gravel, or any soil of a firm character. In other cases, when the subsoil yields readily to the ordinary pressure that the road-surface must bear, a bed of brushwood from 9 to 18 in. in thickness, must be formed to receive the

Fig. 25.

soil on which the road-covering is to rest. The brushwood should be carefully selected from the long straight slender shoots of the branches or undergrowth, and be tied up in bundles, termed *fascines*, from 9 to 12 in. in diameter, and from 10 to 20 ft. long. The fascines are laid in alternate layers crosswise and lengthwise, and the layers are either connected by pickets, or else the withes, with which the fascines are bound, are cut to allow the brushwood to form a uniform and compact bed.

This method of securing a good bed for structures on a weak wet soil has been long practised in Holland, and experience has fully tested its excellence.

CHAPTER V.

ON REPAIRING AND IMPROVING ROADS.

The improvement of existing roads may be divided into two distinct branches; namely, the improvement of their general course and levels, and the improvement of the materials of the road. The first of these consists in the application of the principles which have been laid down for the construction of new roads, and consists generally in straightening their course by extinguishing unnecessary curves and bends; improving their levels by either avoiding or cutting down hills, and embanking valleys; increasing their width, where requisite, and rendering it uniform throughout.

With regard to the improvement of the surface, the operations may consist in reducing the transverse section to the form shown in Fig. 22, page 75, filling up all ruts, cleansing and deepening, if necessary, the side ditches, cutting down trees or hedges by the side of the road, removing mud-banks which but too often exist on the road-sides, and placing proper materials on its surface.

Of all these, the most important to explain, because the most difficult to effect, and the least generally understood, is the method by which the condition of the surface of the road may be improved. In the practice of many surveyors, the remedy for a bad road is to heap on fresh material; whereas, as Mr. Macadam has very justly observed,*
"Generally, the roads of the kingdom contain a supply

* Parliamentary Report, 1819, p. 21.

of materials sufficient for their use for several years, if they were properly lifted and applied." Generally speaking, the cause of inferiority is the imperfect transverse form of the roads, and the improper manner in which the road-materials are used. The remedial measures to be adopted must, in a great degree, depend upon the nature of the materials composing the upper surface of the road; but, whatever these may be, the road must be brought to the proper form of section before much improvement can be expected. This should be done by cutting down those parts which are too high, and raising the depressed parts. But where the surface of the road is so rotten or brittle that the materials lifted are not fit to be again used, the renewal may be done gradually, and rather by the addition of fresh material to the lowest parts. Unless the materials of which the road has been formed are found to be brittle or rotten, or to be already very thin, the course to be pursued is that which is technically termed "lifting the road," which consists in loosening and turning the surface of the road to a depth of about 4 in., and carefully removing such portions of the materials as may be found in an improper state; such as large stones, which should be broken into pieces of the proper dimensions, and then restored to the road. Where, however, the materials of the road are of such a nature that in lifting they would crumble or fall to powder, a different mode of proceeding must be adopted: the surface of the road must be carefully cleared of mud and dirt, and fresh material, prepared as described in the preceding chapter, should be laid on in a very thin coat, never exceeding at one time 3 inches, and, under ordinary circumstances, not more than 2 inches in thickness. Where the surface of a road, although hard, is found to be very thin, it is necessary, instead of lifting the old materials, to add a fresh coat; and, preparatory to doing this, it is well to loosen the surface of the road with a pick, so that

ON REPAIRING AND IMPROVING ROADS. 81

the new material may become more rapidly incorporated with the old material.

Autumn is the best season of the year for repairing roads, when they are in a wet state, for the depressed and soft parts of the road are then not only the most readily detected, but the surface of the road is softer, and the new materials are more easily worked into it. As was observed, the quantity laid on at one time should never exceed 3 inches in depth; and, generally speaking, a half, or even a third, of this thickness would suffice, if judiciously employed. It is certain that roads are more frequently spoilt by having too much material put upon them than by having too little. On this subject Mr. Penfold, whose experience in road-making cannot be questioned, remarks:—"It is one of the greatest mistakes in road-making that can be committed, to lay on thick coats of materials, and when understood, it will no longer be resorted to. If there be substance enough already in the road, which, indeed, should always be carefully kept up, it will never be right to put on more than a stone's thickness at a time. A cubic yard, nicely prepared and broken, as before described, to a rod superficial, will be quite enough for a coat, and will be found to last as long as double the quantity put on unprepared and in thick layers. There is no grinding to pieces when so applied; the angles are preserved, and the material is out of sight and incorporated in a very little time. Each stone becomes fixed directly, and keeps its place: thereby escaping the wear and fretting which occur in the other case."*

Although autumn is the best time of the year for repairing roads, it is not to be assumed that it is the only season for executing repairs. Roads should, so to speak, be always under repair: every road should be divided into lengths, on each of which an intelligent labourer, who

* "A Practical Treatise on the best Mode of Making and Repairing Roads," p. 15.

thoroughly understands his business, should be placed, to attend constantly and at all times to the state of the road, for which he should be held responsible. His duty should consist in keeping the road always scraped clean and free from mud, in filling any ruts or hollow places, the moment they make their appearance, with broken stones. A supply of broken stones should be kept in depôts or recesses formed at the sides of the road at intervals of a quarter of a mile. The depôts should be capable of containing about 30 cubic yards of material, and should be formed with walls, so that the quantity of material in them can be easily measured. Each man should be provided with a wheelbarrow, a shovel, a pickaxe, and a scraper. As autumn approaches, additional labourers should be engaged; but the constant labourers should alone be responsible for the good order of the road.

Not only should the mud formed in wet weather be carefully scraped off from the surface to the sides, and removed altogether as soon as it becomes sufficiently solid, but in dry weather the roads should be constantly and regularly watered. After a long season of drought, the surface of a road becomes, as it were, baked: and in this state, being brittle, it is quickly injured and worn to dust by constant traffic. But a regular and moderate supply of water entirely obviates this undue wear, and preserves the road in a proper state. Care should be taken that the water is properly applied, as much injury may be done by the water being discharged in too great quantity, or unevenly distributed. The manner in which the water should be poured upon the road should resemble, as nearly as possible, a gentle shower of rain. The system of watering roads in particular conditions, even in winter, has been practised with advantage, as is shown by the following extract from the evidence of Mr. Benjamin Farey, surveyor of the Whitechapel Road, before the Committee of the

House of Commons:—" The wheels stick to the materials, in certain states of the road, in spring and autumn, when it is between wet and dry, particularly in heavy foggy weather, and after a frost; by which sticking of the wheels the Whitechapel Road is often, in a short time, dreadfully torn up and loosened; and it is for remedying this evil that I have, for more than eight years past, occasionally watered the road in winter. As soon as the sticking and tearing-up of the materials is observed to have commenced, several water-carts are employed upon these parts of the road, to wet the loamy and glutinous matters so much that they will no longer adhere to the tire of the wheels, and to allow the wheels and the feet of the horses to force down and again fasten the gravel-stones; the traffic, in the course of from four to twenty hours after watering, forms such a sludge on the surface as can be easily raked off by wooden scrapers, which is performed as quickly as possible; after which, the road is hard and smooth. The advantages of this practice of occasional winter-watering have been great; and it might, I am of opinion, be adopted with like advantages on the other entrances into London, or wherever else the traffic is great, and the gravel-stones are observed to be torn up by sticking to the wheels."*

Proceeding with a description of the tools or implements employed in the construction and repair of roads, the most important of these is the level or plummet-rule used for forming the true transverse section of the road. It is shown in Fig. 26, and consists of a horizontal straightedge or bar A C, having in the centre of its length a plummet B D, for ascertaining when the straightedge is horizontal. Thus far it exactly resembles an ordinary bricklayer's level. A line is drawn near the end A of the bar, and at every 4 feet from this line a gauge (a, b, c, d,) is fixed in a

* Parliamentary Report, 1819, p. 40.

Fig. 26.—Plummet Rules.

dovetailed groove, in such a way as to be capable of being moved up or down, so as to adjust the depth of its lower end below the horizontal line of the bottom of the straight-edge; and there are thumb screws (one of which is shown on an enlarged scale in Fig. 27) passing through each gauge, by tightening which the gauge can be fixed when so adjusted. When the bottoms of the gauges a, b, c, and d, have been adjusted as shown in Fig 26, they will coincide with the surface recommended to be given to a road 30 feet in width, and such as is shown in Fig. 22; and, in order to ascertain whether the surface of any existing road is constructed to the proper inclination and form, it is only requisite to apply the level, which, when placed perfectly horizontal, by means of the plummet B D, should rest upon the road at the lower extremity of each of the gauges, *a, b, c,* and *d*. For forming the sides of roads of greater width than 30 feet, it would be convenient to have a level constructed in the manner shown in Fig. 28, in which a straightedge about 15 feet long, has a plummet at the centre of its length, so adjusted that when hanging truly in its place, the lower

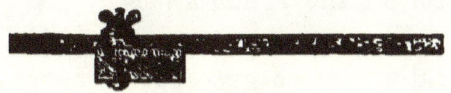

Fig. 27.—Gauge.

side of the straightedge should be inclined from a horizontal line at the rate of 1 in 30.

The *pick* used for lifting the surface of roads is shown in Fig. 29. The bent iron head (*a b*) should weigh about ten pounds, having a large eye in the centre (*c*), in which is fitted the handle, which should be of ash, rather more than 2 feet in length; one extremity (*a*) should be formed

Fig. 28.—Plummet Rule.

like the end of a chisel, while the other (*b*) should terminate in a blunt point. Both ends should be tipped with steel.

The most useful form of *shovel* for road purposes is shown in Fig. 30. The blade should be somewhat pointed, and the handle bent, so as to enable the person using it to bring the blade flat upon the surface of the road without excessive stooping.

The ordinary *wheelbarrows* are of ash or elm, with cast-iron wheels; but they may be made of wrought iron, which would combine strength and durability with lightness. Of whatever material they are constructed, they should not exceed 9 inches in depth, and their sides should be splayed with a slope of 2 to 1. It is also desirable to have hooks placed on their sides to receive a shovel and a pick.

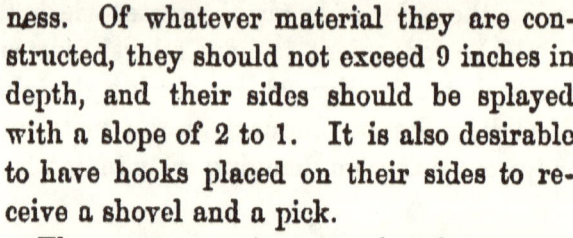

Fig. 29.—Pick.

The *screens*, or *sieves*, employed for separating coarse gravel from hoggin or small gravel, consist of iron wires or slender rods, placed at equal distances apart, and fixed in a frame of wood, the sides of which are raised about 5 inches above the plane of the wires. In the screens the frames are rectangular, about 5 feet 6 inches in height and 3 feet wide, and the wires are stretched in the direction of its length at distances varying from ½ inch to 1¼ inch, according to the size of the stone required; and these wires are kept in place by others crossing them at intervals at 5 or 6 inches. When used, they are placed so that the plane of the wires is inclined about 30° from the upright, and the gravel to be screened being dashed or thrown forcibly against them, the finer particles pass through and fall on the further side of the screen, while the large stones roll down its surface and fall on the nearest side. The sieves are somewhat different in form: the frame is circular, forming a cylinder about 6 inches in depth and 20 inches in diameter, and the wires placed either as already described or equally close in both directions, forming a kind of bottom to the cylinder. The sieve is held hori-

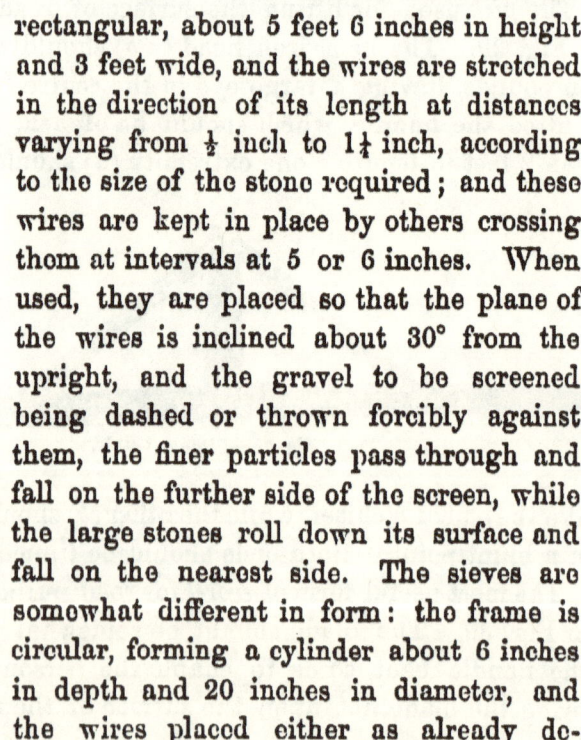

Fig. 30.—Shovel.

zontally by one man, while the other throws into it a

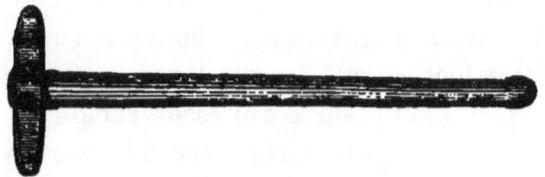

Fig. 31.—Hammer.

shovelful of gravel; upon shaking the sieve, the fine hoggin falls through, leaving the stones in the sieve, which are then thrown by the man into anything which may be placed to receive them. This is generally the best and cheapest mode of screening gravel.

The *hammers* generally employed for breaking stones are of two sizes, and are shown in Figs. 31 and 32. The handles should be of straight-grained ash, and the iron heads of the weight and form shown in the drawings; the faces should be spherical, and case-hardened or steeled.

Fig. 33 represents the ring to be used for testing the size of the broken stones. Its internal diameter is 2¼ inches, and the largest stones should in all positions be passable through the ring.

Fig. 34 represents a *pronged fork*, to be used instead of a shovel for taking up the stones to throw upon the road. The advantages attending its use are, that a man can take up the stones much more quickly and more easily than with a shovel, free from dirt and extraneous matter.

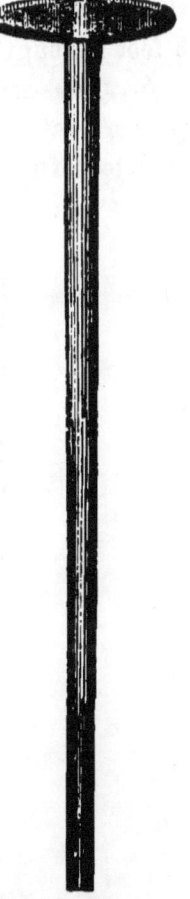

Fig. 32.—Hammer.

It is advantageous to roll the surface of new roads, in

order to consolidate the material; and for this purpose a cast-iron roller is usually employed, about 5 feet wide, 4 feet in diameter, and weighing about 4 tons.

The *rakes*, which should be employed in filling in ruts and hollow places in the surface of roads, should be formed with prongs between 2 and 3 inches in length, fixed at the distance of three-quarters of an inch apart, into a wooden head about 11 inches in length. The handles should be formed of ash, and should be about 6 feet in length.

Fig. 33.—Ring-gauge.

Scrapers are indispensable for preserving roads in a proper state and free from mud. They are usually constructed of wood shod with wrought iron; but it is better

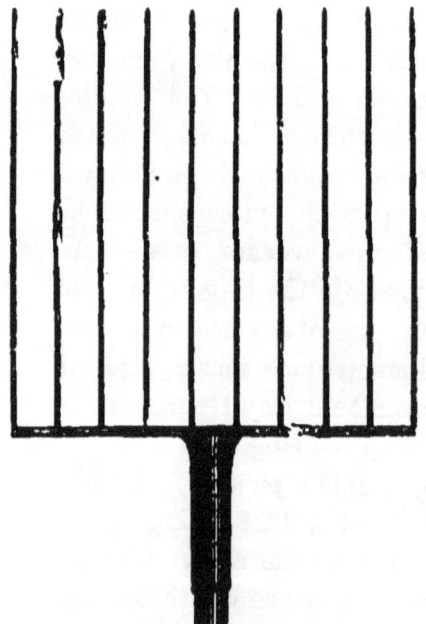

Fig. 34.—Pronged Fork.

to make them entirely of iron. They should be 6 inches in depth, and about 18 inches in length, and slightly curved at each extremity to prevent the escape of mud at each side.

Scraping machines have been invented, and are very generally employed, by means of which the surface of a road may be scraped much more regularly and quickly than with the old scrapers. They consist of a number of iron scrapers, attached to a frame mounted on wheels, which are so placed that, when the body of the machine is raised somewhat, the wheels are lifted from the ground, and the whole weight of the machine is thrown upon the scrapers, which, upon the machine being drawn across the road, scrape all the mud from its surface, and carry it to the sides.

A machine has also been invented by Mr. [now Sir Joseph] Whitworth, of Manchester, which has been extensively employed, both there and in London, for sweeping up the mud from the roads and carrying it away at once. It consists of a species of endless broom, passing round rollers attached to a mud cart, and so connected by cogged wheels with the wheels of the cart that, when the latter is drawn forwards, the broom is caused to revolve, and sweeps the mud from the surface of the road, up an inclined plane, into the cart. The machine is drawn by one horse; and, by its aid, the roads are swept much more rapidly and better than by the old system of scraping, and with less injury to the surface of the road, and less annoyance to the passengers.

CHAPTER VI.

CONSTRUCTION OF ROADS.—FOUNDATION AND SUPERSTRUCTURE.

It has been maintained, by Mr. Macadam and others, that a soft and yielding foundation for a road is better than one which is firm and unyielding. Mr. Macadam has stated that he "should rather prefer a soft one to a hard one;" and even a bog, "if it was not such a bog as would not allow a man to walk over it."* The principles upon which this opinion was founded were, that the road on the soft foundation being more yielding or elastic, the materials of which the covering of the road was formed would be less likely to be crushed and worn away by the passage of a heavy traffic over them than when placed on a hard solid. The contrary opinion is, however, maintained by the largest number of advocates; and it is certain that there is no more general cause of bad roads than soft foundations. A firm, solid, and dry substratum is necessary for the road materials to rest upon. However good the materials themselves may be, and however much care may be bestowed upon the distribution of them, the material and labour are of no avail unless a good foundation has been prepared. The outer surface of the road should be regarded merely as a covering to protect the actual working road beneath, which should be sufficiently firm and substantial to support the whole of the traffic to be carried over it. The proper function of the road materials is to protect the actual road from

* Parliamentary Report, 1819, p. 23.

being worn and injured by horses' feet and wheels, or by the action of the weather. This lower, or *sub-road*, as it may be called, being once properly constructed, would last for ever, provided that the outer case or covering is renewed from time to time, so as to maintain a sufficient depth for the protection of the sub-road.

Roads may be classed as follows, according to the manner in which the foundation is formed:—

1st. Roads having no artificial foundation, but in which the covering materials are laid on the surface of the ground.

2nd. Roads having a foundation of concrete.

3rd. Roads having a paved foundation.

And each of these might be sub-divided according to the kind of material employed as a covering.

The first class of roads comprises by far the largest proportion of the roads in this country. But it should only be employed in cases where the road is not sufficiently important to warrant any large expenditure, and where the anticipated amount of traffic is small.

Every care should be taken to make the road as solid as possible. If the ground is at all of a soft or wet nature, deep ditches should be cut on each side of the line of the road, and cross under-drains should be formed in the manner already described at page 76. And where the ground is very soft, a layer of faggots or brushwood, from 4 to 6 inches in depth, should be laid over the surface of the ground to receive the road materials. For embankments, or in other situations where the ground has been recently deposited, the surface should be either rolled or *punned*, that is, beaten with heavy beetles, so as to confer upon it as great a degree of solidity as possible. The same mode of proceeding should be followed, even where it is intended to form either a paved or a concrete foundation.

The employment of concrete composed of gravel and

lime was first proposed by Mr. Thomas Hughes, and the following remarks upon its use are quoted from his work on roads:—*

"The use of lime concrete, although an introduction of modern times, and certainly one of rather a novel character, derives its real origin from a very remote period. We have indisputable evidence that the Romans, in constructing their military ways, particularly in France, adopted the practice of forming a concrete foundation composed of gravel and lime, on which also they placed large stones as a pavement. The consequence of a construction so solid has been, that, in many parts of Europe, the original bed or crust of the Roman roads is not at the present day entirely worn down, even after a lapse of fifteen centuries.

"With the view of affording a modern example in which lime concrete has been used, I would refer to the Brixton Road, where a concrete composed of gravel and lime has been recently applied by Mr. Charles Penfold, surveyor to the trust. In this case the proportion of gravel to lime is that of four to one. The lime is obtained from Merstham or Dorking, and before being used it is thoroughly ground to powder. The concrete is made on the surface of the road, and great care is taken, when the water is added, that every particle of the lime is properly slaked and saturated. The bed of concrete having been spread to the depth of 6 inches over the half breadth of the road, the surface is then covered over with 6 inches of good hard gravel or broken stone, and this depth is laid on in two courses, of 3 inches at a time, the first course being frequently laid on a few hours after the concrete has been placed on the road. The carriages, however, are not on any account allowed to pass over it until the concrete has become sufficiently hard and solid to carry the traffic without suffering the road material to sink and be pressed into the body of concrete. On the

* "The Practice of Making and Repairing Roads," p. 44.

other hand, the covering of gravel is always laid on before the concrete has become quite hard, in order to admit of a more perfect binding and junction between the two beds than would take place if the concrete were suffered to become hard before laying on the first covering. The beneficial effect arising from the practice of laying on the gravel exactly at the proper time is, that the lower stones, pressed by their own weight, and by those above them, sink partially into the concrete, and thus remain fixed in a matrix, from which they could not easily be dislodged. The lower pebbles being thus fixed, and their rolling motion consequently prevented, an immediate tendency to bind is communicated to the rest of the material—a fact which must be evident, if we consider that the state called binding, or rather that produced by the *binding*, is nothing more than the solidity arising from the complete fixing and wedging of every part of the covering, so that the pebbles no longer possess the power of moving about and rubbing against each other. It is found that, in a very few days after the first layer has been run upon, the other, or top covering, may be applied; and, shortly afterwards, the concrete, and the whole body of road material, becomes perfectly solid from top to bottom. The contrast thus presented to the length of time and trouble required to effect the binding of road materials where the whole mass is laid on loose is alone a very strong recommendation in favour of the concrete.

"The experiment of using concrete on the Brixton Road, although not at present on a very extensive scale, has been tried under circumstances very far from being favourable, and on a part of the road which had hitherto baffled every attempt to make it solid. Since the concrete has been laid down, however, there is not a firmer piece of road in the whole trust."

Mr. Penfold gives the result of an experiment made by

him upon the Walworth Road. "It was raised by nine inches of concrete, and six of granite and Kentish ragstone mixed; and in some parts it was covered by rag and flints. The improvement is so great, with respect to the draught, and so desirable with respect to the saving in the annual repair, that the trust have directed it to be applied to upwards of two miles of road upon which the greatest traffic exists."*

One of the principal advantages attending the employment of concrete as a foundation for roads is, that a good and solid road may be made with materials, such as round pebbly gravel, which, on any other mode of application, would be ill-suited to the purpose, and would form a very imperfect road. This description of gravel is that which is by far the most frequently met with. The gravel selected for this purpose should be free from any kind of dirt, clay, or other impurity, and should consist of stones and sand, mixed in about such proportions that the latter would just fill the interstices of the former. The gravel should then be mixed with the proper quantity of ground unslaked lime—in ordinary cases five or six parts of gravel and one of lime will be found to answer; after which, sufficient water being added to effect the slaking of the lime, the whole should be quickly, but thoroughly, mixed up, and then immediately thrown into place, and trimmed off at once to the proper form intended to be given to its upper surface; the first layer of broken stones or screened gravel should then, as Mr. Hughes directs, be placed just as the concrete is about to set.

The third mode of forming an artificial foundation was introduced by Mr. Telford. It consists in laying a rough pavement on the top of the formation-surface, which is afterwards covered by the road materials. The following

---

* "Practical Treatise on the best Mode of Making and Repairing Roads," by Charles Penfold, p. 31.

is an extract from one of Mr. Telford's specifications for a portion of the Holyhead Road:—" Upon the level bed prepared for the road materials, a bottom course or layer of stones is to be set by hand, in form of a close, firm pavement; the stones set in the middle of the road are to be 7 inches in depth: at 9 feet from the centre, 5 inches; at 12 from the centre, 4 inches; and at 15 feet, 3 inches. They are to be set on their broadest edges lengthwise across the road, and the breadth of the upper edge is not to exceed 4 inches, in any case. All the irregularities of the upper part of the said pavement are to be broken off by the hammer, and all the interstices to be filled with stone chips, firmly wedged or packed by hand, with a light hammer; so that when the whole pavement is finished, there shall be a convexity of 4 inches in the breadth of 15 feet from the centre." *

The stone which Telford employed for this purpose, was generally such as would have been totally unfit for most other purposes, whether on account of its inferior quality, or the smallness of its dimensions. In comparing the two methods of forming artificial foundations of roads, regard must be had to the nature of the materials found in the locality in which the road is to be formed. Where stone is plentiful, and easily procured, the paved foundation would be the best; whilst, where stone is scarce, and gravel and lime abundant, the preference must be given to the concrete foundation.

The foundation of the road having been prepared, the next proceeding is, to lay a firm and compact covering upon the foundation, to form a smooth surface for carriages to travel upon. The materials of which the covering is composed should possess the property of becoming quickly united into one solid mass, of which the surface should be smooth and hard, and at the same time not liable to

* Sir H. Parnell on Roads, p. 133.

be broken to pieces, or ground into dust, by wheels or horses' feet. The materials which have been employed for this purpose are of two kinds; angular fragments of broken stone of different sorts, and gravelly pebbles. It is essential to the formation of a good road that the distinction here pointed out be always kept clearly in view, because a totally different mode of proceeding must be adopted to form a perfect road with these two classes of material. The want of attention to the distinction here pointed out has led to much discussion and misapprehension on the employment of clay, chalk, or other material, as a binding upon roads.

If the materials of which the covering is to be formed are in angular masses, no binding of any description is requisite; as they quickly become united by dovetailing, as it were, amongst each other—much more firmly than they would by the use of any kind of artificial cement.

When the stones, instead of being angular, are round and pebbly, like gravel stones, it is necessary to mix with them just sufficient foreign matter, of a binding nature, as will serve to fill up the interstices between the stones, for otherwise these would roll about, and would prevent the road from becoming solid.

There are, then, two methods of cementing or solidifying the surface of a road: one, by the mechanical form of the materials themselves forming a species of bond; the other, by the use of some cementing or binding matter. And in comparing the relative merits of the two, the preference must certainly be given to the former, that in which the stones are united in virtue of their angular form, without the use of any cementing material. The principal reason for this preference is, that roads formed with stones so united, are not materially affected by wet or frosty weather; whereas, roads whose surfaces are composed of pebbly stones united by cementing material, become loose and rotten under such circumstances : the cementing material becoming softened by the wet, and reduced to a loose pulverulent condition by subsequent frost.

The first method, that of forming the road-covering entirely with angular pieces of stone, without any other material, was first strongly recommended by Mr. Macadam, and subsequent experience has shown its superiority over every other which has been employed. The most important quality in stone for road-making is *toughness* : mere hardness without toughness is of no use, as such stone becomes rapidly reduced to powder by the action of wheels. Those stones which have been found to answer this purpose best are, the whinstones, basalts, granites, and beach pebbles. The softer descriptions of stone, such as the sandstones, are not fitted for this purpose, being far too weak to resist the crushing action of wheels. The harder and more compact limestones may be employed; but, generally speaking, the limestones are to be avoided, in consequence of their great affinity for water, by which in frosty weather, which has been preceded by wet, they are split up into powder, when the solidity of the road is destroyed.

Next in importance to the quality of the stone is its preparation. This consists in reducing it to angular fragments of such a size that they will pass freely through a ring of $2\frac{1}{2}$ inches in diameter, in every direction; that is, that their largest dimensions shall not exceed that measure.

The stone thus prepared, should be evenly spread over the surface prepared for the foundation of the road, to a depth of about 6 inches; and the road should then be opened for traffic. In Mr. Telford's specifications, he usually directed that, on the top of this coating of broken stone, a layer of good clean gravel, about $1\frac{1}{4}$ inch in depth, should be spread before throwing the road open for use. The reason for this practice was, to lessen the extreme unevenness of the surface, and to render the road more pleasant to pass over when first opened. It would be better, however, for the public to put up with the temporary inconvenience of a rough road, because the gravel does a permanent injury to the road, and reduces in a considerable degree the facility with which the stones unite into a compact, solid mass.

Broken stone, being superior to gravel for the purpose of road-making, should always be employed where it can be easily obtained. The quality of gravel varies so considerably, that while some kinds may, when properly prepared, form a very excellent road, others may be entirely worthless; such as those kinds of gravel the stones composing which are of the sandstones and flints, for even flints, although hard, are so excessively brittle as to be immediately crushed by the passing of the wheels over them. The gravel when taken from the pit should be passed over a screen which will allow all stones less than $\frac{1}{4}$ of an inch to pass through it, and the fine stuff, or *hoggin*, as it is technically termed, thus obtained, should be reserved for forming the footpaths; the remainder, which has not passed through the screen, should have all the stones whose greatest dimension is more than $2\frac{1}{2}$ inches removed and broken, and it would be desirable that these broken stones should be reserved for the upper layer. In screening the gravel, especially as it comes from the pit, a certain portion of loam is generally found adhering to the stones, and this should by no means be separated from them, for, although angular broken stones require no extraneous substance to cause them to bind, the pebbles of which most gravel is composed, require a certain amount of loam, clay, or chalk, to fill up the interstices between the stones, and to prevent them from being rolled about, as they otherwise would be. The gravel thus prepared by screening should be laid on and spread to a uniform depth of not more than 6 inches over the whole surface. The road thus finished is thrown open to the public.

On this subject, Mr. Hughes says:*—"In laying on this upper covering many surveyors commit a great error in not making a distinct difference between angular or broken stones and those rounded smooth pebbles of which gravel is usually composed. The former cannot be too well cleaned before being laid on the road, because, even when entirely divested of all earthy matter, they soon become wedged and bound closely together when the pressure of carriages comes upon them. But the case is different with the smooth, round surfaces of gravel; for if this material be entirely cleaned by means of washing and repeated siftings, the pebbles will never bind, until in a great measure they become ground and worn down by the constant pressure and rubbing against each other. Before this takes place, the surface of the road must be considerably weakened, and will, in fact, be incapable of supporting the pressure of heavy wheels, which consequently sink into it, and meet with considerable resistance to their progress. Under these circumstances, it seems that the practice of too scrupulously cleaning the rounded pebbles of gravel must be decidedly condemned; and the question then arises, to what extent should the cleaning process be dispensed with? or, in other words, what proportion of the binding material found in the rough gravel, as taken out of the pit, should be allowed to remain in the mass intended to be placed on the road? * * * A long course of experience, accompanied by attentive observations on these details in the practice of road-making, has convinced me that it is much better and safer, as a general rule, to leave too much of the binding material in the

* "The Practice of Making and Repairing Roads," p. 15.

gravel than to divest it too completely of this substance. When the gravel is placed on a road without being sufficiently cleaned, the constant wear and tear, aided by the occurrence of wet weather, causes the harder material or actual gravel to be pressed close together; and the surplus of soft binding material remaining after the interstices between the pebbles are filled up, being then forced to the top, and usually mixed with water, becomes mud, and according to the usual practice should be scraped to the sides of the road. When this has been done, the surface is usually firm and solid; because the hard gravel below the mud has become perfectly bound, without, at the same time, being broken or ground to pieces. Suppose, next, a road covered with gravel, too much cleaned, where it is evident that the destruction of the gravel will continue until it becomes broken into angular pieces, and a sufficient quantity of pulverised material has been formed to hold the stones in their places and thus to effect the binding of the mass. I need hardly say, that the deterioration thus occasioned to the road is an evil of much more importance, and one much more to be avoided, than that occasioned by employing stones not sufficiently cleaned. Regardless of all this, however, it is the practice of many road-surveyors to insist that all gravel, of whatever quality, shall be rendered perfectly clean by repeated siftings, and even by washing, until it becomes entirely divested of all that may properly be considered the binding part of the material."

Particular care and attention is required to be given to new roads when opened for traffic; a sufficient number of men should be employed to keep every rut raked in the moment it appears, and guards or fenders should be placed on the road, in order that vehicles may be caused to pass over every part of the surface in turn. If these precautions are not taken, years may elapse before the road attains a firm condition; and many roads have been permanently ruined through want of sufficient attention when first used. When ruts are once formed, vehicles using the road keep in the same track, deepening and increasing the rut. In wet weather, the rut is filled with water, which, having no other means of escape, slowly penetrates the sides and bottom of the rut, rendering them so soft as to be more susceptible to the action of every succeeding carriage. Ruts thus formed, involve a much larger outlay to repair the injury than that which would have been sufficient to prevent it; besides the inconvenience, danger, and expense to the public, in travelling on a road in such a condition.

Amongst the substances which were mentioned as binding material to be mixed with *clean* gravel, was chalk. Many roads have been ruined by its improper use. There are two modes in which chalk may be advantageously employed in the construction of roads. It may be laid in the *very bottom* of the road, to form the foundation; but it must be at such a depth as to be entirely beyond the influence of frost, otherwise it will quickly destroy the road, for chalk has a very powerful affinity for water; or rather, to speak more correctly, capillary attraction for it, in consequence of which it readily absorbs all the moisture which finds its way through the road covering. Herein consists the value of chalk if judiciously applied, for the water thus absorbed would otherwise penetrate to the foundation of the road and render it soft. If, however, the chalk be placed within the reach of frost, the water, which is only mechanically held by the chalk, will, in the act of congealing, expand, and by so doing rend the chalk into fragments, and reduce it, in fact, to a pulverulent state, in which condition it is changed by the succeeding thaw into a soft paste or mud. The other purpose for which chalk may be employed is, as before mentioned, to be mixed with gravel in order to make it bind; though, in using chalk for this purpose, it is only required when the gravel is perfectly clean and free from other binding matter. The mixing of chalk with gravel already containing sufficient clay or loam is not only useless, but is positively injurious; and, even when the gravel is of such a nature as to require the intermixture of chalk, care should be taken not to add too much chalk, for it is not with chalk as with the loam or clay, with which gravel is naturally combined. Clay, generally speaking, possesses little power of absorbing water, but the superabundant chalk would soon be reduced to the state of a soft paste by the action of the weather, in the manner which has been described. Chalk, therefore, if used as a binding material with gravel, on the surface of roads, should be reduced to a state of powder, and should be perfectly and thoroughly mixed with the gravel before the latter is spread on the road.

Although the use of bushes or bundles of faggots has been recommended for the foundation of roads over very soft or boggy ground, they should only be employed in such a situation, and they should be placed at such a depth below the surface, as will ensure their always being damp; for when they are placed where they would be alternately wet and dry, they quickly become rotten, and form a soft stratum beneath the road.

NOTE BY THE EDITOR.—Mr. Walker,* in 1819, speaking of the great advantage of filling up or grouting the joints of granite pavement with lime-water, which finds its way into the gravel between and under the stones, and forms the whole into a concrete mass, made a suggestion of value, in recommending, for the same purpose, a mixture of a

* Evidence before the Select Committee on the Highways of the Kingdom, 1819.

little of the borings or clippings of iron, or small scraps of hoop-iron, with the gravel used in filling up the joints of the paving. The water would, he said, very soon create an oxide of iron, and form the gravel into a species of rock. "I have seen," he adds, "a piece of rusty hoop taken from under water, to which the gravel had so connected itself, for 4 or 5 inches round the hoop, as not to be separated without a small blow of a hammer. And the cast-iron pipes which are laid in moist gravel, soon exhibit the same tendency."

CHAPTER VII.

PAVED ROADS AND STREETS.

For roads or streets, through towns or cities where the traffic is considerable, a paved surface is preferable to a macadamised surface. Macadamised roads in such situations are exposed to incessant and heavy traffic, their surface is rapidly worn, and it requires constant repair, and much attention in scraping or sweeping, or in raking-in ruts. The cost of maintenance is very heavy, whilst there are frequent interruptions of, or interference with, traffic. In dry weather, the macadamised road is dusty, and in wet weather it is covered with mud. The only advantage which such a road really possesses over a pavement consists in the less noise produced by carriages in passing over it.

Several different methods have been employed of forming the foundations of pavements, such as concrete and broken stone; but where it can be done, it is perhaps best to lay the new pavement on the old surface of the road, whether paved or macadamised, taking care, of course, that its surface has been brought first to an even state, and of the required form of cross section.

The practice of laying the new pavement on the top of the old has been a great deal used in Paris, and has there been found to answer extremely well. It is usual to take up and relay the old pavement, in order that its surface may be even and true; after which it is covered with gravel, on which the new stones are bedded. Mr. Telford strongly recommended the surface upon which the pavement was intended to be laid to be prepared as though intended for a macadamised road, and that it should be used, in that state, by carriages until it had become thoroughly consolidated; when the pavement should be laid on the top of the hard road so formed, the stones being properly bedded in a kind of coarse mortar. Mr. Edgeworth, in his work on roads, states that this method of forming paved roads had been extensively employed in Dublin, and was attended with considerable success.

In constructing a paved road or street, the following method should be employed for forming the foundation. The loose ground at the surface should first be entirely removed. The depth to which it may be requisite to do this depends upon the nature of the ground; and, unless the ground be very solid, it should be removed to such a depth as will allow of 18 inches of concrete beneath the pavement. In some situations so great a thickness as this may not be requisite; but it is better to err rather in forming too strong a foundation, than in forming one the failure of which necessitates the taking up and relaying of the pavement. When the loose ground has been removed, a layer of concrete, prepared in the manner already described at page 92, should be evenly spread over the whole area of the intended road. The depth of concrete should never be less than 12 inches; and, under ordinary circumstances, it should not be less than 18 inches. Its upper surface should have the true form of cross section intended to be given to the road.

The stone sets should be well bedded upon the concrete in a kind of coarse mortar, which should also be well filled in between their joints. For the sets, several of the harder kinds of stone are used, such as granite, whinstone, and the very hardest varieties of limestone and freestone. Of all materials, granite is the best for sets, and more particularly those kinds, such as the Guernsey and Aberdeen, which do not wear smooth and acquire a polished surface. Smooth wear is a great practical objection attending the employment of excessively hard stone. With regard to the form of the stones, experience has shown that the best form is that of rectangular blocks, from 7 to 9 inches in depth, depending on the amount and nature of the traffic, and not more than 3 or 4 inches in width. Until lately, it was considered better to have stones of much greater width, under the supposition that, having a larger base, they would be better able to support the superincumbent weight, but experience has shown that the narrow stones are much the better. The stones should be sorted according to their depths and widths: for, if the stones are of unequal depth, and the surface of the concrete has been made, as it should be, even and parallel to the intended surface of the road, any stones less than the general depth would require more mortar to be placed under them, and would consequently settle down more than the others, and form hollows on the surface of the road. They should also be sorted according to their width, so that they may run entirely across the street in parallel courses, as shown in Fig. 35, and the stones in each course should be so selected as to break joint with those in each adjoining course, as there shown.

Fig. 35.—Plan of Pavement.

A firm and substantial curb should be laid on each side of the road, to act as an abutment; and, in laying the pavement, the courses should be commenced at each side, and worked towards the middle; the joints between the stones of each course should be as thin as possible, and the last stone should fit tightly, so as to form a kind of key to the course. After the stones have been set they should be well rammed down with a heavy punner, and stones which sink below the general level should be taken up, and re-packed underneath. It is not usual to incur the expense of bedding the stones in mortar, in the manner here recommended. The ordinary practice is to pour a thin grouting of sand and lime over the surface, after the pavement has been laid, which finds its way, although very imperfectly, into the interstices between the stones. But, this is mistaken economy, for a pavement laid as here described, upon a firm concrete foundation, and having the joints perfectly formed with good mortar, would last almost for ever, since nothing else than the positive wearing away and destruction of the stone would render its renewal necessary.

Paved roads should be watched for a few months after having been opened for traffic, in order to prevent irregular settlement in the stones, and consequently an uneven and irregular surface. When any portion of the road is found to settle below the general level, it should be taken up, and a sufficient quantity of fine concrete put underneath it to bring it slightly above the level of the surrounding surface. Great attention should be given to the manner in which the pavement is re-laid after having been disturbed for the repair of sewers, water-pipes, or gas-pipes: the excavated ground, when thrown back, should be well punned, or beaten down in layers of not more than a foot in thickness, and at least 18 inches of concrete should be laid on the top, under the pavement. The surface of the fresh concrete should be about an inch above the general level of the concrete, and the stones should be properly bedded in the manner already described, care being taken that the stones correspond in depth and width with those already in place on either side, and, furthermore, that the last stone, in making good each course, fits tightly into its place.

In laying pavement in streets having a considerable

inclination, two methods have been employed to afford a more sure and perfect hold for the horses' feet than the ordinary pavement. The first method is shown in Fig. 36, and consists in laying between the rows of paving-stones

Fig. 36.
Pavement on Inclines.

a course of slate, rather less than an inch in thickness, and about an inch less in depth than the stones. By this means a series of small channels or grooves, about an inch in width and depth, are formed between the rows of stones, which affords sufficient stay for the horses' feet. The other method is somewhat simpler, and consists in merely placing the ordinary paving-stones somewhat canted on their beds, as shown in Fig. 37, so as to form a series of ledges or steps, against which the horses' feet being planted, a secure footing is obtained.

Fig. 37.
Pavement on Inclines.

Roadways and streets in cities must be accompanied by sidewalks, and crossing-places, for foot-passengers. The sidewalks are made of large flat flag-stones, 3 inches thick, laid on the ordinary ground, or on gravel, well rammed and settled. The proper width of the sidewalks depends on the amount of the traffic. It would, in all cases, be well to have them at least 12 feet wide; they are laid at a slope, or pitch, of 1 inch to 10 feet, towards the pavement, to convey the surface-water to the side channels. The pavement is separated from the sidewalk by a row of long slabs, termed curb-stones, which confine both the flagging and the paving-stones. They are usually about 12 inches by from 6 to 10 inches in section, and they are laid, in some cases, on their side, in other cases on edge. They form the sides of the side channels, and they should for this purpose project a few inches above the outside paving-stones, and be sunk at least 4 inches below their upper surface. The curb-stones are flush with the upper surface of the sidewalks, to allow water to run over into the side channels.

The crossings should be from 4 to 6 feet wide, and slightly raised above the general surface of the pavement, to keep them free from mud.

CHAPTER VIII.

ON HEDGES AND FENCES.

In most situations, fences are required to mark the boundaries of roads, and separate them from the adjoining lands. They should be dispensed with wherever it is possible, for the reason that fences, of whatever kind, deprive, to a greater or less extent, the surface of the road of the benefit of free exposure to the action of wind and sunshine, both of which are essential for maintaining it in a dry state. Few persons are aware of the extent to which a road may be injured by high hedges, or lines of trees. Trees are worse than hedges, because they not only deprive the road of the action of the air and sun, but they further injure it by the dripping of rain from their leaves, as a consequence of which the road is kept in a wet state long after it would otherwise have become dry.

When fences are indispensable, they should be placed as far as may be from the sides of the road, and should be kept as low as possible. When there is a deep ditch on either side of the road, it becomes necessary, to prevent accident, that the fence should be placed between the road and the ditch; but, in other situations, the fence should be placed on the field-side of the ditch. In so doing, the surface-draining of the road into the side ditches is less interfered with, and the action of air and sunshine is less obstructed by the fence.

The different descriptions of fence which may be employed are various. In districts where stone is plentiful, and especially in the immediate neighbourhood of quarries, where stone rubble can be obtained at a trifling cost, dry rubble walls, without any mortar, are very good and cheap, and require little or no repair.

For the road itself, an open post-and-rail fence is the best which can be employed, because it scarcely impedes the action of the wind and the sun upon the surface of the road; but the great practical objection to timber fences is, their liability to decay, which occasions frequent and constant expense for renewal.

The most common, and, all things considered, the most useful, fence is the quickset hedge. If properly planted, and carefully attended to for the first few years, a natural fence may thus be obtained, sufficiently strong to resist the efforts of cattle to break through, and very economical in cost for maintenance. A bank or mound of earth, at least two feet in depth, should be prepared for the reception of the quicks, which should be three-years plants which have been transplanted two years. The best kind of soil is one of a light sandy nature, admitting sufficient moisture to nourish the plants, and retaining moisture in dry seasons. Heavy clay soils are not sufficiently pervious to water, and plants placed in such soils are never found to thrive. A mixture of peat or of rotten leaves is of great use, and causes the plants to grow with much vigour. The quicks are most commonly planted in a single row, at distances of about 4 inches apart. But a much better hedge is formed by planting them 6 inches apart, in two double rows, as shown in Fig. 38, with a space of 6 inches between the rows,

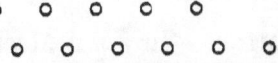

Fig. 38.—Planting Hedges.

and so arranged that the plants in one row are opposite

the spaces in the other. By this arrangement, although the plants are really not so crowded, and have more space round their roots from which to derive nourishment than in the single row, they form a thicker hedge. The proper time for planting quicks is during the autumn or the spring, and, in fine seasons, the operation may be continued during the whole winter. A temporary fence should be put up to protect the young plants from injury; and the fence should be retained until the hedge has attained sufficient strength to require its protection no longer: at the end of a period, under favourable circumstances, of three or four years after the quicks are planted. That the plants may thrive, they must be very carefully attended to at first, and it is essential that they should be properly cleaned and weeded at least twice every year. Once every year, towards the end of the summer, the hedge should be judiciously trimmed, not to such an extent as to produce stunted plants, but by merely cutting off the upper and more straggling shoots, so as to bring it to a level and even surface. By proceeding in this manner, a neat, strong, and compact hedge of healthy plants will be obtained in about three years after planting.

When the hedge or fence is placed between the road and the side ditches, it is essential that small drains be formed at least every fifteen yards, to convey the water from the side tables or gutters, through the fence, into the ditches.

Professor Mahan insists that fences and hedges should not be higher than 5 feet; and that no trees should be suffered to stand on the road-side of the side drains, for, independently of shading the roadway, their roots would in time throw up the road-covering.

NOTE BY THE EDITOR.—Sir John Macneil proved experimentally, on the Holyhead Road, the unfavourable influence of close and high hedging in interfering with what may be called the ventilation of a road:—keeping it moist, and incurring excessive draught. By means of his experiments on tractional resistance, the trustees and the surveyors of the roads "have," he said, "perceived the defective parts of the road; and within three months after the Report of the Parliamentary Commissioners became public, there was not a hedge on that part of the road where the draught was shown to be excessive, that was not cut down, and improved on the surface."*

Mr. Walker had previously, in 1819, remarked that nothing is more injurious to roads than the permitting of high hedges and plantations near them.†

* Report of the Select Committee on Steam Carriages, 1831; p. 103.
† Report of the Select Committee on the Highways of the Kingdom, 1819.

CHAPTER IX.

ON TAKING OUT QUANTITIES FOR ESTIMATES.

THE process of making out an estimate for any description of engineering work may be divided into two distinct parts; namely, in the first place, calculating the actual quantity of each description of work to be executed; and, in the second place, affixing to these quantities just and reasonable prices, such as the work might really be executed for.

In the construction of roads, the principal item of expense is the earthwork, or the cost of forming the cuttings and embankments to obtain the required levels for the formation service, in excavating the ditches and forming the banks, and in laying on the metalling or ballast to form the road. Of these different descriptions of work, the first, namely, the cuttings and embankments, are the only ones the estimation of the quantity of which is attended with any difficulty. The others, being generally constant, are readily obtained by ascertaining the quantity in a given length, as a yard, and then multiplying that quantity by the total length of the road.

In the calculations of solid contents required in balancing the excavations and embankments, the most accurate method consists in subdividing the different solids into others of the most simple geometrical forms, as prisms, prismoids, wedges, and pyramids, whose solidities are readily determined by the ordinary rules for the mensuration of solids. As this process, however, is frequently long and tedious, other methods requiring less time, but not so accurate, are generally preferred, as their results give an approximation sufficiently near the true for most practical purposes. They consist in taking a number of equidistant profiles, and calculating the solid contents between each pair, either by multiplying the half sum of their areas by the distance between them, or else by taking the profile at the middle point between each pair, and multiplying its area by the same length as before. The latter method is the more expeditious; it gives less than the true solid contents, but a nearer approximation than the former, which gives more than the true solid contents, whatever may be the form of the ground between each pair of cross profiles.

In calculating the solid contents, allowance must be made for the difference in bulk between the different kinds of earth when occupying their natural bed and when made into embankment. From some careful experiments on this point made by Mr. Elwood Morris, published in the *Journal of the Franklin Institute*, it appears that light sandy earth occupies the same space both in excavation and embankment; clayey earth about one-tenth less in embankment than in its natural bed; gravelly earth about one-twelfth less; rock in large fragments about five-twelfths more, and in small fragments about six-tenths more.

The calculation may often be simplified by considering the slopes apart from the trunk or main body of the cutting. For example, let Fig. 39 be the section of a cutting, A D being the natural surface of the ground; then if B C is the width of the formation surface, B C F E will be the trunk or central portion of the cutting, and A E B, F C D, will be the side slopes. Now, the width of the first is constant, being the formation width, while its depth varies as the depth of the cutting; and therefore its cubic content for a given length forward is directly proportional to the depth of the cutting. If, for instance, the line G H were the natural surface of the ground instead of A D, the cutting being now twice as deep as before, the cubic content of the trunk I B C K would be twice as great as that of E B C F.

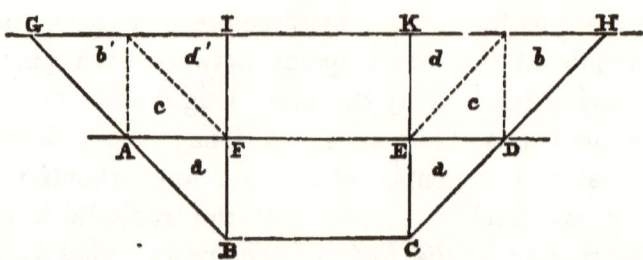

Fig. 39.—Section of a Cutting.

But the cubic content of the slopes increases in the proportion of the square of the depth of the cutting, as is evident from the figure, in which I B being *twice* as great as E B, the volume of the slope G I B is *four* times as great as A E B, the triangles *a*, *b*, *c*, and *d* being evidently equal. In consequence of these two quantities varying in a different proportion, it is convenient to estimate them separately. In order to facilitate the calculation of earthwork, several tables have been published, the most complete and elaborate of which are those by Sir John Macneil. The following table, No. 6, will be found very useful in estimating the content of cuttings or of embankments of moderate depth, and will give the result very nearly true, except in cases in which the two ends of the cutting are of very different depths. The first, fourth, and seventh columns contain the depth of the cutting or height of embankment, in feet, for every tenth of a foot, from 0·1 to 25·2 feet; the second, fifth, and eighth columns express the content, in cubic yards, of one foot in width, and one chain in length, of a portion of the trunk or central part of a cutting, whose mean depth is shown in the preceding column. The quantities taken from these columns must be multiplied by the formation width. The third, sixth, and ninth columns express the content in cubic yards of a length of one chain of both slopes, when the slopes are formed at 1 to 1. With any other ratio the quantities derived from these columns must be increased in the same proportion. Thus, if the slopes are 3 to 1, the quantity obtained from the table must be multiplied by 3.

TABLE No. 6.—CONTENTS OF CUTTINGS OR OF EMBANKMENTS.

Height or depth.	Content for a length of one chain.		Height or depth.	Content for a length of one chain.		Height or depth.	Content for a length of one chain.	
	Of each foot in width of trunk.	Of slopes taken at 1 to 1.		Of each foot in width of trunk.	Of slopes taken at 1 to 1.		Of each foot in width of trunk.	Of slopes taken at 1 to 1.
Feet.	Cub. yds.	Cub. yds.	Feet.	Cub. yds.	Cub. yds.	Feet.	Cub. yds.	Cub. yds.
0·1	·24	·02	4·3	10·51	45·20	8·5	20·78	176.6
0·2	·49	·10	4·4	10·76	47·32	8·6	21·02	180·8
0·3	·73	·22	4·5	11·00	49·50	8·7	21·27	185·0
0·4	·99	·39	4·6	11·24	51·72	8·8	21·50	189·3
0·5	1·22	·61	4·7	11·49	54·10	8·9	21·76	193·6
0·6	1·47	·88	4·8	11·73	56·42	9·0	22·00	198·0
0·7	1·71	1·20	4·9	11·98	58·79	9·1	22·24	202·4
0·8	1·96	1·56	5·0	12·22	61·11	9·2	22·49	206·9
0·9	2·20	1·98	5·1	12·47	63·58	9·3	22·73	211·4
1·0	2·44	2·44	5·2	12·71	66·10	9·4	22·98	216·0
1·1	2·69	2·96	5·3	12·96	68·66	9·5	23·22	220·6
1·2	2·93	3·52	5·4	13·20	71·28	9·6	23·47	225·3
1·3	3·18	4·13	5·5	13·44	73·94	9·7	23·71	230·0
1·4	3·42	4·79	5·6	13·69	76·66	9·8	23·96	234·8
1·5	3·67	5·50	5·7	13·93	79·42	9·9	24·20	239·6
1·6	3·91	6·26	5·8	14·18	82·23	10·0	24·44	244·4
1·7	4·16	7·06	5·9	14·42	85·09	10·1	24·69	249·4
1·8	4·40	7·92	6·0	14·67	88·00	10·2	24·93	254·3
1·9	4·64	8·82	6·1	14·91	90·96	10·3	25·18	259·3
2·0	4·89	9·78	6·2	15·16	93·96	10·4	25·42	264·4
2·1	5·13	10·78	6·3	15·40	97·02	10·5	25·67	269·5
2·2	5·38	11·83	6·4	15·64	100·1	10·6	25·91	274·6
2·3	5·62	12·93	6·5	15·89	103·3	10·7	26·16	279·9
2·4	5·87	14·08	6·6	16·13	106·5	10·8	26·40	285·1
2·5	6·11	15·28	6·7	16·38	109·7	10·9	26·64	290·4
2·6	6·36	16·52	6·8	16·62	113·0	11·0	26·89	295·8
2·7	6·60	17·82	6·9	16·87	116·4	11·1	27·13	301·2
2·8	6·84	19·16	7·0	17·11	119·8	11·2	27·38	306·6
2·9	7·09	20·56	7·1	17·36	123·2	11·3	27·62	312·1
3·0	7·33	22·00	7·2	17·60	126·7	11·4	27·87	317·7
3·1	7·58	23·49	7·3	17·84	130·3	11·5	28·11	323·3
3·2	7·82	25·03	7·4	18·09	133·8	11·6	28·36	328·9
3·3	8·07	26·62	7·5	18·33	137·5	11·7	28·60	334·6
3·4	8·31	28·26	7·6	18·58	141·2	11·8	28·84	340·4
3·5	8·56	29·94	7·7	18·82	144·9	11·9	29·09	346·2
3·6	8·80	31·68	7·8	19·07	148·7	12·0	29·33	352·0
3·7	9·04	33·46	7·9	19·30	152·6	12·1	29·58	357·9
3·8	9·29	35·30	8·0	19·56	156·4	12·2	29·82	363·8
3·9	9·53	37·18	8·1	19·80	160·4	12·3	30·07	369·8
4·0	9·78	39·11	8·2	20·04	164·4	12·4	30·31	375·9
4·1	10·02	41·09	8·3	20·29	168·4	12·5	30·56	381·9
4·2	10·27	43·12	8·4	20·53	172·5	12·6	30·80	388·1

Height or depth.	Content for a length of one chain.		Height or depth.	Content for a length of one chain.		Height or depth.	Content for a length of one chain.	
	Of each foot in width of trunk.	Of slopes taken at 1 to 1.		Of each foot in width of trunk.	Of slopes taken at 1 to 1.		Of each foot in width of trunk.	Of slopes taken at 1 to 1.
Feet.	Cub. yds.	Cub. yds.	Feet.	Cub. yds.	Cub. yds.	Feet.	Cub. yds.	Cub. yds.
12·7	31·04	394·3	16·9	41·31	698·2	21·1	51·58	1088
12·8	31·29	400·5	17·0	41·56	706·4	21·2	51·82	1099
12·9	31·53	406·8	17·1	41·80	714·8	21·3	52·07	1109
13·0	31·78	413·1	17·2	42·04	723·2	21·4	52·31	1119
13·1	32·02	419·5	17·3	42·29	731·6	21·5	52·56	1130
13·2	32·27	425·9	17·4	42·53	740·1	21·6	52·80	1140
13·3	32·51	432·4	17·5	42·78	748·6	21·7	53·04	1151
13·4	32·76	438·9	17·6	43·02	757·2	21·8	53·29	1162
13·5	33·00	445·5	17·7	43·27	765·8	21·9	53·53	1172
13·6	33·24	452·1	17·8	43·51	774·5	22·0	53·78	1183
13·7	33·49	458·8	17·9	43·76	783·2	22·1	54·02	1194
13·8	33·73	465·5	18·0	45·00	792·0	22·2	54·27	1205
13·9	33·98	472·3	18·1	44·24	800·8	22·3	54·51	1216
14·0	34·22	479·1	18·2	44·49	809·7	22·4	54·76	1227
14·1	34·47	486·0	18·3	44·73	818·6	22·5	55·00	1238
14·2	34·71	492·9	18·4	44·98	827·6	22·6	55·24	1249
14·3	34·96	499·9	18·5	45·22	836·6	22·7	55·49	1260
14·4	35·20	506·9	18·6	45·47	845·7	22·8	55·73	1271
14·5	35·44	513·9	18·7	45·71	854·8	22·9	55·98	1282
14·6	35·69	521·1	18·8	45·96	864·0	23·0	56·22	1293
14·7	35·93	528·2	18·9	46·20	873·2	23·1	56·47	1305
14·8	36·18	535·4	19·0	46·44	882·4	23·2	56·71	1316
14·9	36·42	542·7	19·1	46·69	891·8	23·3	56·96	1327
15·0	36·67	550·0	19·2	46·93	901·1	23·4	57·20	1339
15·1	36·91	557·4	19·3	47·18	910·5	23·5	57·44	1350
15·2	37·16	564·8	19·4	47·42	920·0	23·6	57·69	1362
15·3	37·40	572·2	19·5	47·67	929·5	23·7	57·93	1373
15·4	37·64	579·7	19·6	47·91	939·1	23·8	58·18	1385
15·5	37·89	587·3	19·7	48·16	948·7	23·9	58·42	1397
15·6	38·13	594·9	19·8	48·40	958·3	24·0	58·67	1408
15·7	38·38	602·5	19·9	48·64	968·0	24·1	58·91	1420
15·8	38·62	610·2	20·0	48·89	977·8	24·2	59·16	1432
15·9	38·87	618·0	20·1	49·13	987·6	24·3	59·40	1443
16·0	39·11	625·8	20·2	49·38	997·4	24·4	59·64	1455
16·1	39·36	633·6	20·3	49·62	1007	24·5	59·89	1467
16·2	39·60	641·5	20·4	49·87	1017	24·6	60·13	1479
16·3	39·84	649·5	20·5	50·11	1027	24·7	60·38	1491
16·4	40·09	657·5	20·6	50·36	1037	24·8	60·62	1503
16·4	40·33	665·5	20·7	50·60	1047	24·9	60·87	1516
16·6	40·58	673·6	20·8	50·84	1058	25·0	61·11	1527
16·7	40·82	681·7	20·9	51·09	1068	25·1	61·36	1540
16·8	41·07	689·9	21·0	51·33	1078	25·2	61·60	1552

As an example of the use of this table, we may estimate the quantities in the cutting and embankment shown on the working section, Fig. 20, page 39. In the following table, the first column contains the number of the peg, the second the depth of cutting or height of embankment, the third the cubic content of the corresponding portion of the trunk, and the fourth column the content of the slopes:—

Cutting No. 1.

No. of peg.	Depth of cutting.	Trunk.	Slopes.
1	·0	·37	·11
2	·6	1.47	·88
3	·9	2·20	1·93
4	1·2	2·93	3·52
5	1·6	3·91	6·26
6	1·9	4·64	8·82
7	2·0	4·89	9·78
8	1·9	4·64	8·82
9	1·9	4·64	8·82
10	2·1	5·13	10·78
11	2·2	5·38	11·83
12	2·4	5·87	14·08
13	2·3	5·62	12·93
14	2·3	5·62	12·93
15	2·5	6·11	15·28
16	2·1	5·13	10·78
17	2·3	5·62	12·93
18	2·5	6·11	15·28
19	2·2	5·33	11·83
20	1·6	3·91	6·26
21	·8	1·96	1·56
		91·53	185·46
		40	
		3661.20	
		185·46	
		3846·66 cub. yds.	

Embankment No. 1.

No. of peg.	Height of embankment.	Trunk.	Slopes.
22	·6	1·47	·88
23	2·2	5·38	11.83
24	3·5	8·56	29·94
25	4·0	9·78	39·11
26	3·8	9·29	35·30
27	2·6	6·36	16·52
28	1·3	3·18	4.13
29	·3	·73	·22
		44·75	137·93
		40	
		1790·00	
		137·93	
		68 96	
		1996·89 cub. yds.	

ESTIMATES.

It will be remarked that, at peg 1, where the cutting has no depth, we have yet inserted quantities in the third and fourth columns. The manner in which these were derived is as follows. Although at the peg itself there is no cutting, at the next peg the depth is 0·6, and therefore the *mean depth* is 0·3, which, in the table, gives 0·73 and 0·22. As, however, we have here only half a chain (as it is the commencement of the section), we insert half these quantities, or 0·37 and 0·11. The sum of all the separate portions of the trunk, being obtained by addition, is then multiplied by 40, the width of the formation, to which the sum of the slopes being added, gives 3846·66 cubic yards as the content of the cutting. In the case of the embankment, the slopes being 1½ to 1, we add *one and a half* times the sum of the slopes, and thus obtain 1996·89 cubic yards as its content. By reference to the working section, Fig. 20, it will be seen that the quantities there given agree with the above.

PART II.

RECENT PRACTICE IN THE CONSTRUCTION OF ROADS AND STREETS.

BY D. K. CLARK, C.E.

CHAPTER I.

MATERIALS EMPLOYED IN THE CONSTRUCTION OF ROADS AND STREETS.

For carriage-ways: Stones.—The hardest and toughest kinds of stones are those which are employed in the construction of paved and macadamised carriage-ways. First in geological order, as well as in the order of usefulness, is granite.

Granite is an unstratified or igneous rock, generally found inferior to or associated with the oldest of the unstratified rocks, and sometimes penetrating them in the form of dykes and veins. It is a crystalline compound of three simple minerals — felspar, silica or quartz, and mica; in which the proportion of silica varies from 65 to 80 per cent. Granite derives its name from its coarse *granular* structure—*granum*, Latin for grain. The silica generally occurs in an amorphous condition, enclosing and cementing together the felspar and the mica in the state of crystals. Some granites are ternary compounds; others are quaternary, or even quinary compounds, consisting of silica, two varieties of felspar, or two varieties of mica.

Two felspars are present in some of the granites of Galway, Donegal, and Aberdeenshire. Two micas occur in the Wicklow granite, imparting black and grey hues. Instead of mica, another substance, hornblende, is found in some granites. Such granites, of which the component parts are quartz, felspar, and hornblende, are distinguished as *syenite*, or syenitic granites; so called, because it was first found in the island of Syene, in Egypt.

Of the constituents of granite, *quartz*, glassy in appearance, is a compound of the metallic base silicium and oxygen; *felspar*, opaque, is of a yellowish or a pink colour, composed of silicious and aluminous matter, with a small proportion of lime and potash; *mica* consists principally of clay and flint, with a little magnesia and oxide of iron; *hornblende* is a dark green crystalline substance, composed of flint, alumina, magnesia, with a considerable proportion of black oxide of iron.

The granite districts of the United Kingdom, whence most of the material for road-making is obtained, are in Devon, Cornwall, the Channel Islands of Guernsey and Herm; Mount Sorrel, in Leicestershire; Clee Hills, in Shropshire; Port-Nant and Portmadoc, in North Wales; Aberdeen, in Scotland; and Newry, in Ireland. The syenitic granites of Guernsey, Herm, Mount Sorrel, and North Wales, are celebrated for their hardness and durability. The granite of Devon and Cornwall is of several varieties, and it has a prevailing greyish hue. Aberdeen granite is of a bluish-grey tint; so also is the Newry granite. Mount Sorrel granite is of a rich pink colour.

Granites vary much in hardness and resistance to crushing force. In some places, by the decomposition of the felspar or the mica, they are sufficiently soft to admit of their being dug out with a spade. The specific gravity of granite varies from 2·60 to 3·00, and the volume of 1 ton is from 12 to 14 cubic feet. Conversely, the weight of a cubic yard of solid granite varies from 1·93 to 2·25 tons, averaging, say, 2 tons; whilst the average weight of a solid cubic foot is 1½ cwt. Granite is capable of absorbing, on an average, a gallon, or 10 lbs., of water per cubic yard, or 1-450th of its weight of water.

Mr. Mallet found that the maximum resistance of 1-inch cubes of various granites to crushing force, ranged from 1 to 6 tons. The annexed table, No. 7, gives the results of experiments to determine the resistance of granites:—*

TABLE No. 7.—CRUSHING RESISTANCE OF GRANITES.

Locality of Stone.	Surface exposed to pressure.	Pressure per square inch.	
		To fracture.	To crush.
	Square in.	Square in.	Square in.
Herm	16	4·77	6·64
Aberdeen (Blue)	17½	4·13	4·64
Heytor	16	3·94	6 19
Dartmoor	16	3·52	5·48
Peterhead (Red)	18	2·88	4·88
Peterhead (Bluish grey)	18	2·86	4·36
Penrhyn	16	2·58	3·45
Killiney (Grey felspathic)	9	—	4·81
Ballyknocken (Coarse grey)	9	—	1·43
Ballybeg, Carlow (Grey felspathic)	9	—	3·17

Some results of experiments made by Sir William Fairbairn† on the crushing strength of building stones are added, in Table No. 8. The specimens were, for the most part, 2-inch cubes:—

* The data for this table are derived from Sir John Burgoyne's "Rudimentary Treatise on Blasting and Quarrying," 1862, page 94; except for the last three stones, for which the data are borrowed from Professor Hull.

† "Memoirs of the Literary and Philosophical Society of Manchester," 1857; vol. xiv. page 31.

TABLE No. 8.—CRUSHING STRENGTH AND ABSORBENT POWER OF
VARIOUS STONES.
(Deduced from Sir W. Fairbairn's data.)

Stone.	Locality of Quarry.	Specific gravity.	Cubic feet in one ton.	Crushing resistance per square inch.	Weight of water absorbed after 48 hours' immersion.
			Cub. ft.	Tons.	1 part in
Porphyry	France	—	—	18·04	—
Greywacké	Penmacnmaur	2·75	13·04	7·54	1641
Do.	Ingleton	2·79	12·87	—	1963
Granite	Aberdeen	—	—	5·16	—
Do.	Mount Sorrel	2·67	13·45	5·74	490
Do.	Bonaw, Inverary	—	—	4·87	—
Sandstone	Yorkshire	2·41	14·85	4·38	47·5
Do.	Runcorn	—	—	·98	—
Limestone	—	—	—	3·80	—
Do. Magnesian	Auston, Worksop	—	—	2·26	—
Brick, hard	—	—	—	·84	—
Do. red	—	—	—	·36	—

The *Trap Rocks* are a large group of igneous rocks allied to granite, composed of felspar, augite, and hornblende. The various proportions and states of aggregation of these simple minerals, and their differences in external forms, give rise to varieties known as Basalt, Dolorite, Greenstone, Whinstone, Greywacké, and others. The term *trap* is derived from *trappa*, a Swedish word for stair, because the rocks of this class sometimes occur in large tabular masses, rising one above another like steps. Basalt is one of the most common varieties; it is a dark green or black stone, composed of augite and felspar, very compact in texture, of considerable hardness. It often contains iron; whence the name *basalt*, an Ethiopian word for iron. Dolorite is likewise composed of augite and felspar. Greenstone is composed of hornblende, which is dark green, and felspar. Greywacké is a very hard rock, as heavy as granite, and much harder. The rock is very often of a grey colour; whence the name. *Wacké* is a provincial miner's term in Germany.

The principal uses of the trap rocks are for paving and macadamising. The most important quarries are those of Barden Hill, in Leicestershire, and of Penmaenmaur, in North Wales. The greywacké of Penmaenmaur is manufactured into squared sets, which are sent by ship and by rail in enormous quantities to Manchester, Liverpool, and other large towns in the north of England and in Scotland, and to Ireland.

Comparative Wear of Stones.—Mr. Walker tested the comparative durability of various granites and one whinstone. In his first experiments, he submitted square specimens of Guernsey, Aberdeen, and Peterhead granites, having equal weights and rubbing surfaces, to frictional motion on a large block of stone, with sand and water between them. The Guernsey granite showed the smallest amount of wear.

He made further tests for the durability of granites and whinstone, under working conditions, by laying down, in 1830, two parallel lines of rectangular tram-stones, 18 inches wide and 12 inches deep, in the gateway of the Limehouse turnpike, so as to be exposed to all the heavy traffic from the East and West India Docks. The experimental stones were all new. After an exposure of 17 months to the wear and tear of the traffic, the stones were taken up, and the loss by wear was found to be in order as follows:—

TABLE No. 9.—RELATIVE WEAR OF GRANITES, ETC.
(MR. WALKER.)

Name of Stone.	Superficial area.	Loss of weight per sq. foot.	Vertical wear.	
	Sq. foot.	pounds.	inch.	relative.
Guernsey granite	4·73	·95	·060	1·000
Herm granite	5·25	1·05	·075	1·190
Budle whinstone	6·34	1·22	·082	1·316
Peterhead blue granite	3·48	1·80	·131	2·080
Heytor granite	4·31	1·92	·141	2·238
Aberdeen red granite	5·38	2·14	·159	2·524
Dartmoor granite	4·50	2·78	·207	3·285
Aberdeen blue granite	4·82	3·06	·225	3·571

From the table, it is apparent that Guernsey granite is many times tougher and more durable than Aberdeen granite. Mr. Walker made a still further test of the relative wear of these two stones on old Blackfriars Bridge, which was paved, in 1840, with these granites, in sets 3 inches wide by 9 inches deep, on a bed of concrete 12 inches thick. When the pavement had been down 13 years, it was found that the Aberdeen stone was worn down to the extent of $1\frac{1}{2}$ inch of vertical wear, whilst the Guernsey stone had only worn down $\frac{1}{4}$ inch—showing that the Guernsey had worn only one-sixth as much as the Aberdeen granite.*

The durability of the Welsh stones—the greywacké of Penmaenmaur, and the syenites of Portmadoc, &c., is almost beyond estimation. After 20 years of work in the streets of Manchester, Penmaenmaur stones, the resistance of which to crushing is considerably greater than that of granite, have suffered little more wear than a rounding of the upper surface. But they are extremely slippery, and they make a dangerous pavement. The Irish, the syenitic, and the blue Aberdeen granites retain rough wearing surfaces; but, in consequence of their roughness, they are not so durable as the Guernsey and the Penmaenmaur stones. The rough-wearing sets are excellent stones, and they are specially suitable for the paving of streets on steep inclines.

For macadam, the hardest stones—as Guernsey granite, and Penmaenmaur greywacké—are the most suitable. Their slippery qualities are of no moment in macadam, whilst their hardness and toughness are valuable qualities.

In India, the stones used for macadam are granite, trap, and the hard limestones and sandstones. Laterite, which is a hard sandstone, is very much used on the Madras roads; but it is comparatively soft, and does not bear much traffic.

* Proceedings of the Institution of Civil Engineers, 1853-54; vol. xiii. page 237.

Kunkur is the material chiefly used in Hindostan: it is a peculiar formation of oolitic limestone, found generally in nodules, sometimes in masses a little below the surface of the earth. It makes an excellent road, but it requires constant repair if the traffic is heavy.

For Footpaths.—The flagstones usually employed for footpaths are of sandstone. The Old Red Sandstone of Scotland produces the well-known sandstones and flags of Arbroath and Dundee, which are largely used in Glasgow, Edinburgh, London, and other large towns. The Arbroath pavement is of a light greenish, grey colour, and of dense structure; it resists the weather successfully. The flagstones of Caithness, Cromarty, and Nairn are also used for paving. The carboniferous sandstones are largely used for paving in the north of England and in Scotland; they are generally hard and durable, of yellowish or greyish tints, and of various degrees of coarseness. They are derived from the millstone-grit and the coal measures. Excellent flagstones, which are sent to all parts of England, are obtained from the lower part of the coal formation in Cheshire, near Macclesfield; Lancashire, near Wigan, Burnley, and St. Helen's; and from Elland, in Yorkshire. The sandstones of the coal measures—for instance, those of Heddon and Kenton, near Newcastle—are seldom durable, and they generally become iron-stained on exposure, owing to their containing considerable quantities of iron and alumina.*

The following particulars of the composition of sandstones are derived from the Commissioners' Report of 1839:—

---

* See "Quarries and Building Stones," by Professor Hull; "British Manufacturing Industries."

TABLE No. 10.—COMPOSITION, SPECIFIC GRAVITY, AND STRENGTH OF SANDSTONES.

	Craigleith.	Darley Dale.	Heddon.	Kenton.	Mansfield.
Composition.	per cent.	per cent.	per cent.	per cent.	per cent.
Silica	98·3	96·4	95·1	93·1	49·4
Carbonate of lime	1·1	0·3	0·8	2·0	26·5
Carbonate of magnesia	0·0	0·0	0·0	0·0	16·1
Iron, alumina	0·6	1·3	2·3	4·4	3·2
Water and loss	0·0	1·9	1·8	0·5	4·8
	100·0	100·0	100·0	100·0	100·0
SPECIFIC GRAVITIES.					
Of dry masses	2·232	2·628	2·229	2·247	2·338
Of particles	2·646	2·993	2·643	2·625	2·756
COHESIVE POWERS. In tons per square inch	3·52	3·17	1·78	2·22	2·28

A few particulars of sandstones are given in Table No. 8, page 125.

All sandstones are porous, and they absorb water. The Yorkshire sandstones are capable of absorbing 1 part by weight of water in 48 parts of stone. Sandstone flags are in this respect objectionable, for they retain wet impurities, and remain damp some time after rain.

An instructive test of the comparative durability of various kinds of flagstones was conducted in Liverpool, by Mr. Newlands, in 1856. The stones were found to range, in the order of durability, as follows:—

   Kilrush (Irish), most durable,
   Caithness,
   Knowsley,
   Port Rheuddin,
   Llangollen,
   Yorkshire, least durable.

Of these, Port Rheuddin stone, one of the softest, had the best appearance. Next to this was the Caithness stone. The Kilrush flags had a rough or wrinkled surface; but in the course of time they wore smooth, and their appearance was good. The Llangollen-slate flag looked well, but it was not durable. Caithness flags were extensively substituted for Yorkshire paving; Welsh and Irish flags were also laid to a considerable extent.

In substituting granite curbstones for sandstone curbs, Mr. Newlands formed the wearing surface of the new curbs to the section produced by abrasion upon the softer stone:—thus presenting a larger wearing surface to the strokes of heavy wheels, and minimising the superficial lateral wear.

Asphalte.—Asphalte, the mineral, is a bituminous limestone, or pure carbonate of lime naturally impregnated with bitumen. At the same time, asphalte is, in a scientific sense, synonymous with bitumen.

Bituminous limestone, or asphalte, is a rock consisting of from 90 to 94 per cent. of carbonate of lime, and from 6 to 10 per cent. of bitumen. The rock is of a liver-brown, or deep chocolate colour; it takes an irregular fracture, with definite cleavage, and the texture and the grain vary with the layers. The specific gravity is from 2·20 to 2·30. Exposed to the atmosphere, asphalte gradually assumes a light grey tint, caused by the evaporation of bitumen from the surface.

The asphalte used for the construction of carriage-way pavements, is brought chiefly from the Val de Travers, a few miles from the town of Neuchatel, in Switzerland. Limmer asphalte is brought from a mine situated at Limmer, near the city of Hanover, and from another mine at Vorwohle, near Alfeld, in Brunswick. The Seyssel asphalte is obtained from mines at Pyrimont and Garde Rais, at Seyssel, in France.

Asphalte, as excavated, must be pulverised before being used, by decrepitation at a temperature of from 212° to 300° Fahr., or by mechanical means. It is applied in two modes for the formation of pavements:—1st, as *compressed asphalte*, which is obtained by heating the powder to from 212° to 250° Fahr., and causing the particles to cohere by the application of pressure. 2nd, as *liquid asphalte*, or *asphaltic mastic*, for which a manufacturing process is necessary. The powder is heated with from 5 to 8 per cent. of free bitumen; the bitumen acts as a flux, the mixture melts, and is run into moulds for use, forming cheeses 12 inches in diameter and 4 inches thick. Like rock asphalte, mastic, heated alone, does not melt; it only becomes soft. To be re-melted, a fresh quantity of bitumen must be added. For paving, a large proportion of sand or small gravel is mixed with it, even to the extent of 60 per cent. No chemical union takes place between the mastic and the sand, but the cohesion is so complete that the fracture of sanded mastic shows the simultaneous fracture of the grit; and the power of resistance to the heat of the sun is increased by the admixture.

The Val de Travers and the Seysell rocks are the only two natural asphaltes used in laying compressed-asphalte roadways. The former has not quite so close a grain as the latter, and it is softer, and is more regularly impregnated with bitumen.

Artificial Asphalte.—By mixing heated limestone and gas-tar, a material possessing some of the properties of asphaltic mastic has been obtained. But gas-tar is a very inferior form of bitumen for paving purposes: it passes from the dry to the liquid state, and conversely, according to the season, and it is brittle.

Wood.—Both soft and hard woods are employed for paving: closed-grained wood of the pine family, as Canadian and Baltic yellow pines; and hard woods, as oak, elm, ash, and beech. Spruce and white deals are not very suitable for paving purposes. The following Table,* No. 11, shows the average crushing resistance of cubes of timber, of from 1 inch to 4 inches, lineal dimension, in the direction of the fibre:—

TABLE No. 11.—CRUSHING RESISTANCE OF TIMBER.
Cubes, 1 inch; 2 inch; 3 inch; and 4 inch.

Name of Timber.	Crushing resistance per sq. inch.
	tons.
English oak, unseasoned	2·19
Do. seasoned	3·34
Dantzic oak	3·34
American white oak	2·71
Moulmein teak	2·56
Iron wood	5·21
Greenheart	6·44
Honduras mahogany	2·85
English elm	2.58
Canadian elm	2·45
Dantzic fir	3·10
Riga fir	2·34
Spruce fir	2·17
Larch	2·60
Red pine	2·54
Yellow pine	1·88
Pitch pine	2·88

Mr. D. T. Hope,† in 1844, recorded the results of experiments made by him on the comparative wear and endurance of wood-paving, laid in the same line of carriage-way, resting on the same concrete foundation, and exposed to the same kind and amount of traffic. The blocks of wood were constructed with the fibre placed vertically, placed horizontally, and also placed at various inclinations to the

* Derived from "A Manual of Rules, Tables, and Data," by D. K. Clark, 1877, page 541.
† "Transactions of the Royal Scottish Society of Arts," 1844, page 231.

vertical. Continuous observations were made for a period of 18 months. It was found that the wear or reduction of height of the wood blocks was greater during the first three months than during the following quarters. The more rapid wear at the commencement of the trial was ascribed by Mr. Hope to initial compression of the fibres: the blocks with vertical fibres presented a more compact surface than when they were originally laid, and the surface was so impregnated with sand that it had more of the appearance of sand than of wood. The following shows the total amount of wear at the end of 18 months:—

			Wear at the end of 18 months.
Wood, vertical fibre	.	.	$\cdot$125 inch, or $\frac{1}{8}$ inch
,,	fibre inclined at 75°	.	$\cdot$147 ,, $\frac{8}{50}$,, bare
,,	,,	60	. $\cdot$182 ,, $\frac{3}{16}$,, ,,
,,	,,	45	. $\cdot$241 ,, $\frac{1}{4}$,, ,,
,,	,,	30	. $\cdot$312 ,, $\frac{4}{16}$,, ,,
,,	,,	15	. $\cdot$379 ,, $\frac{3}{8}$,, full
,,	,,	horizontal	$\cdot$480 ,, $\frac{1}{2}$,, bare

It is clear that the resistance of the wood was most effective when the fibres were in a vertical position, and least effective when they were horizontal. The horizontal fibres presented, when worn, the appearance of heaps of broken strings. It is evident, further, that the more nearly the fibres were presented in an upright position, the less was the wear, and, of course, the greater was the durability.

CHAPTER II.

CONSTRUCTION OF MODERN MACADAM ROADS.

In addition to the detailed account of the construction of roads, given by Mr. Law, in the first part of this work, a few words may be added as to the usual modes of construction now adopted. The levels, gauges, plummet-rules, and prong-shovels, formerly employed in the setting out and construction of roads, described at pages 84 to 88, are not now used. "Boning rods" are used for fixing the inclination of the surface, longitudinally and transversely, by the eye.

First-class Metropolitan Roads.—When the ground has been excavated and levelled, it should be rolled, when it consists of clay. A bottoming, or bed, 12 inches thick, of "hardcore," is laid on the ground; it may consist of brick-rubbish, clinker, old broken concrete, broken stone or shivers, or any other hard material, in pieces. The bed should be rolled down to a thickness of 9 inches, and any loose or hollow places made up to the level.

Next comes a layer of Thames ballast, 5 inches thick, rolled solidly to a thickness of 3 inches. The ballast serves to fill up the vacancies in the bottoming, and, being less costly, saves so much of the cost for broken granite.

Broken granite, or macadam, is laid upon the prepared surface of the ballast, in two successive layers, 3 inches thick, rolled successively, to a combined thickness of 4 inches; a layer of sharp sand, $\frac{1}{2}$ inch or $\frac{3}{4}$ inch thick should be scattered over the second layer, and rolled into it with plenty of water. But it is better to add the sand and the water as the second layer of granite is laid, and to roll them well together.

Broken Guernsey granite, which is hard and enduring, is the best material for the coating; Aberdeen granite is softer, and is not so good for the purpose. Flints are, in some instances, laid instead of granite; but they [are worst of all for the coat, for they are brittle, and are soon reduced, by the traffic, to small fragments.

In transverse section, the contour of the road is a segment of a circle. Indeed, all roads and streets are, or should be, circular in section. The general practice is, in this respect, at variance with the practice recommended by Mr. Law, page 75, where he recommends that the section of the surface should be formed of two inclined straight lines, joined by a flat curve at the middle, forming a species of ridge. The advantage of the circular section consists in the fulness given to the "shoulders" of the road, which lie in the lines of the traffic on each side of the centre line of the road, and thus present a full surface for wear.

Second-class Metropolitan Roads are usually constructed of a hard core of brick rubbish, or other material, about 9 inches thick, and a layer of broken granite, or of flints, about four inches thick. The material is not, in general, submitted to rolling, except by the action of the traffic.

Country Roads.—The regular width of country roads is 18 feet; but some of them are only 10 feet wide, and others may be more than 18 feet. The bottom is excavated with a flat floor, not being rounded to the contour of the surface. The bottom layer, or hard core, consists of any hard and dry rubbish:—broken bricks, lumps of chalk, or hard flints. The material most usually employed consists of flint-stones, which are got out of the fields. The largest stones, which may be as much as 6 or 8 inches across, are laid at the bottom. Chalk flints answer best for the bottoming. They are rough, and they bind firmly together, and are open, affording free space for drainage. If the bottom is of solid material, as chalk or rock-stuff, the bed is laid in to a thickness of 12 inches. If it be soft, as of clay or of sand, it is made up to a thickness of 18 inches.

The finishing layer, or coating, consists of flints of a smaller size; they are broken up, if necessary, to the usual size of macadam, and are laid 6 inches thick. No binding of any kind is employed, nor is it the practice to roll the roads.

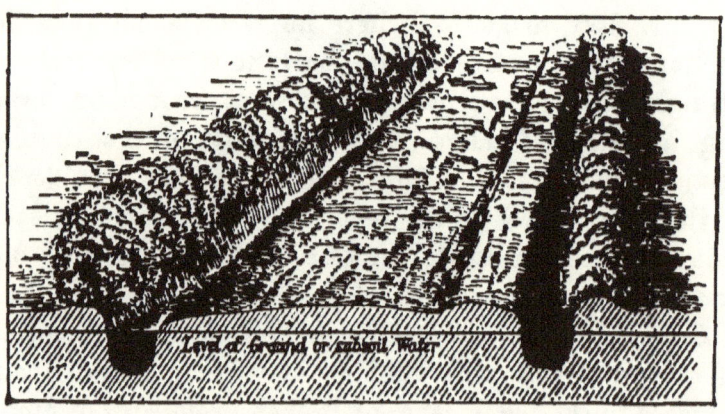

Fig. 40.—Country Road, with open ditches.

The rise of a road, at the middle, is 3 inches for a width of 18 feet. The rule for the rise is to allow 1 inch for every 6 feet of width:—giving a slope of 1 in 36, which is sufficient for running off rain-water, and leaving the roads dry. For roads 10 feet wide, a greater degree of rise is given, where ditches cannot be got in. Openings are made through the fences at each side to let off surface-water into the adjoining fields. Mr. J. G. Crosbie Dawson states that he has made many gravel roads, from 12 to

18 inches thick, at a cost of from 2s. 6d. to 3s. 6d. per square yard.

The general conformation of a country road, with hedges, and with an open ditch at one or both sides, for collecting surface and drainage water, is shown by Fig. 40. Side ditches are a frequent cause of overturns and other accidents. The improved country road, proposed by the *Committee of the Society of Arts,*[*] is represented by Fig. 41,

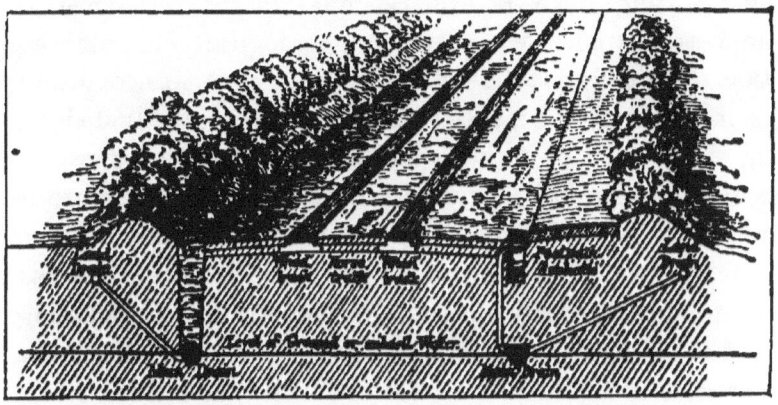

Fig. 41.—Proposed Country Road, with wheel-tracks and subsoil drains.

in which the side drains are suppressed, and replaced by covered drains, which act also as subsoil land-drains. Wheel-tracks of asphalte, horse-tracks of concrete, and a foot-path of asphalte are also proposed.

[*] Report of the Committee of the Society of Arts on *Traction on Roads*, 1875.

CHAPTER III.

MACADAMIZED ROADS.—WEAR.

It was maintained by Macadam (see page 90) that for ease of draft, a yielding, elastic road was preferable to a road constructed according to the prescription of Mr. Telford, on a rigid or paved foundation. The distinction rests entirely upon the precise signification attached to the word "elastic." If the elasticity be perfect, like that of an indiarubber wheel-tire, where the reaction or uplifting force is exactly equal to the resistance to depression under a load, it may be logically maintained that the elastic road is better for draft. But this condition is impossible in practice; and Mr. Provis exactly defined an elastic road as "that which will give way by pressure, and rise again to its original position when the pressure is removed"—tacitly assuming that time is consumed in the reaction, *after* the pressure is removed. Moreover, as Sir John Macneil justly observed, "where there is a yielding and elasticity in the road, there must be motion among the particles with which the road is formed, and this motion produces wear."* And, again: "If the road be weak or elastic, and bend or yield under the pressure of the wheels, the particles of which it is composed will move and rub against each other, or perhaps break by the action of heavy wheels over them."† Herein lies the gist of the

* "Roads," by Sir Henry Parnell, page 412.
† Report of the Sele Committee on Steam Carriages, 1831, page 97.

argument against the admission of any perceptible range of elastic action; and Sir John Macneil touched the essential element of legitimate elastic action, when he located it "on the surface and at the point of contact." This is precisely the mode of elastic action of a wood-pavement:—a material which, with its vertical fibre, acts as a coat of velvet might be supposed to do, simply to absorb and neutralise minute vibration; and it is the only kind of action that is admissible in the work of a road. Sir John Macneil adduced the results of experiments on London streets, in which he found that, when broken granite was laid down, there was little crushing on the surface, but a good deal below. He found "that a great portion of the wear took place near the bottom; the stones there got round after a very short time. They were all jumbled together; the lower part of the stratum was on clay. Before it became a solid mass at all, the wheels worked through, and the stones were kept in motion, and rubbing against each other, from top to bottom." Then he tried the experiment of laying "a portion on a very solid foundation, and the same quantity was put over it. I took up a portion of both roads," he adds, "and we found, where the road was about 15 or 16 inches thick [with a 9-inch pavement], and the stones 6 inches thick over the broken pavement, they were quite square, and as perfect as when put on. In the other case [simple macadam], it was not so." "There are two things to be considered, when a road is newly made: there is very great wear indeed, in the first instance, if there be not a pitched foundation; that is different from the wear that takes place when it is nearly consolidated. To bring it into a solid state, the wear is great. If there are 4 inches of broken stone on the top of a pitched foundation, you may get that road into a perfectly hard and consolidated state, by the ordinary travelling over a turnpike road, in about three months;

but if you put on the same thickness of stone, without a foundation, you will not get it into a perfect state in three or four times as long." *

The bearing of Sir John Macneil's exhaustive evidence is very direct upon the comparative wear of "weak or elastic roads," and roads with a rigid pavement-bottom. He adduces an instance in which an experiment was accurately made to test the relative wear of roads so constructed: "The wear was found to be 4 inches of hard stone, when it was placed on a wet clay bottom; while it was not more than half an inch on a solid dry foundation, or with a pavement bottom, on a part of the same road, where it was subject to the same traffic." †

Estimates have been formed of the relative tear and wear of roads due to the action of the horse-shoes and the action of wheels. Mr. John Farey stated that a cart-horse, walking at a speed of $2\frac{1}{2}$ miles per hour, could draw with a force of traction of 100 lbs., on an average, for 8 hours a day; but that a stage-coach horse, running at a speed of 10 miles an hour, could not exert more than 28 lbs. force of traction, on an average, for $1\frac{1}{2}$ hours a day. The comparative efficiency of the animal is readily worked out. The work done per hour, on duty, and per day, are as follows:—

	Speed in miles per hour.	Traction.	Work done in pounds through one mile.	
			per hour. mile-pounds.	per day. mile-pounds.
Cart-horse	$2\frac{1}{2}$	100 lbs.	250	2000
Stage coach-horse	10	28 ,,	280	420

The comparison clearly indicates that, whilst the work done by a horse per hour at work, is a little greater at 10 miles per hour than at a fourth of the speed, the day's work, on the contrary, is very much less,—little more than

---

* "Roads," by Sir Henry Parnell, pages 412, 413.

† Report of the Select Committee on Steam Carriages, 1831, page 97.

a fifth. It is probable that the injury done to the road is in the same proportion as the day's work done. "In heavy wagons," says Mr. Farey, "drawn slowly by horses, the horses do far more injury, by digging and scraping with their feet, than is done by the horses in coaches and vans travelling quickly; because the wagon-horses, having a heavy pull to make, must choose places in the road where they can place their feet in depressions in order to get hold; hence, on a smooth, good road, they slip, and scrape up the surface." Again: "The horses, by treading with their feet, excavate and scrape out depressions in the surface of the road; that is particularly the case before the road-materials are consolidated into a solid mass. In this manner, the horses, after injuring the road themselves, prepare the way to further injury to the road by the wheels of carriages."*

With respect to the comparative action on a road of the feet of horses and the wheels of vehicles, Mr. Telford considered that the tearing up of a well-made road by horses' feet was much more injurious to the road than the pressure of the wheels.

Sir John Macneil stated that a four-horse stage-coach weighed from 15¾ cwt. to 18 cwt., and frequently carried upwards of 2 tons of passengers and goods;- and the total average weight amounted to from 2¼ to 2½ tons gross. The tires of the wheels were, in general, 2 inches wide; in some instances rather less than 2 inches. Some tires were rounded, forming in cross section a segment of a circle 1¼ inch in diameter. The action of these rounded tires was found to be extremely injurious to the road; coachmen complained that carriages fitted with such rounded wheel-tires "ran wild" in descending hills in summer; and that they ran "heavy" in winter, when the roads were

* Report of Select Committee on Steam Carriages, 1831, page 57.

soft and muddy. Mail-coaches weighed very nearly 20 cwt., and frequently carried a ton of letters and parcels, in addition to their complement of passengers. The tires of their wheels were 2¼ inches wide. Again, the four-horse vans, which travelled at a speed of 6 miles per hour, weighed, on an average, 4¼ tons gross, with 2½-inch tires. There were four descriptions of wagon in general use—8-horse, 6-horse, 4-horse, and farm-wagons of 2 or 4 horses. Sir John Macneil maintained that, on all gravel roads, however made, without foundation or bottoming, the load on the road for a wheel 4 inches wide at the tire should not exceed 15 cwt.; and for wheels of less width, 10 cwt. He gave a table showing the weight actually carried per inch of width of tire-bearing, and what he conceived might be the width of tires if they were made cylindrical, with an even bearing, allowing 1 inch of total width of four wheels for each ton of gross weight; or 1 inch of width in each tire per 5 cwt. of gross weight.

TABLE NO. 12.—WEIGHT OF VEHICLES AND WIDTH OF TIRES ON COMMON ROADS.
(SIR JOHN MACNEIL.)

Description of Vehicle.	Speed in miles per hour.	Average gross weight.	Width of wheels and tires.	Pressure of each wheel.	Pressure per inch of width.	Width of wheel, allowing 5 cwt. per inch of tire.
	miles.	tons.	inches.	cwts.	cwts.	inches.
Mail coach	9 to 11	2	2¼	10	4·40	2
Stage coach	8 to 11	2½	2	12·5	6·25	2½
Van	6 to 7	4¼	2½	21·25	8·29	4¼
Wagon	2½ to 3	6	9	25	2·77	6
Do.	2½ to 3	4½	6	22·5	3·75	4½
Do.	2½ to 3	3½	4	17·5	4·37	3½

From the table it appears that, in the four-horse van, the greatest load per inch of tire upon the road amounted to 8·3 cwt.; and though Sir John Macneil found that good hard roads may support concentrated loads of 10 cwt. per inch of width of tire, he considered that, for "the generality of roads," the concentrated load per inch of width of tire ought not to exceed a maximum of 5 cwt. He judged that the wear of roads caused by horses' feet increased with the speed of transit; and adduced in support of this conclusion the facts that, whilst a coach-horse, going at 10 miles per hour, ran 270 miles per month, and wore away 4 pounds of iron in shoes; a wagon-horse going at 3 miles per hour, and walking 26 miles per day, or 416 miles per month, wore away 4·8 pounds of iron in shoes. From these data, it appears that the coach-horse ran 67·5 miles for each pound of iron worn off, and that the wagon-horse walked 86·5 miles for each pound of iron worn off. These data, though they show a greater rate of wear at the higher speed, scarcely suffice to confirm the conclusion drawn by Sir John Macneil; for the wear of shoes is due entirely to abrasion, whilst that of the surface of the road is due to abrasion combined with excavation. According to another of his estimates, it appears that, in running 63,000 miles, 1,000 pounds of shoe-iron were worn away; showing that 63 miles were run by a horse for each pound of shoe-iron worn off. It also appeared that for an equal mileage of mail-coaches and stage-coaches, of which the hind wheels were 4 feet 8 inches high, and the fore wheels 3 feet high, the tires were worn down from the original section, 2 inches wide by $\frac{3}{4}$ inch thick to a thickness of $\frac{1}{4}$ inch, losing 327 pounds; showing that 193 miles were run by a coach for each pound of tire-iron worn off, or three times the mileage run per pound of shoe-iron.

From the evidence of the comparative wear of shoe-iron and tire-iron, and other considerations, Sir John Macneil framed an estimate of the relative proportions in which a road was deteriorated by the action of atmospheric changes, wheels, and horses' feet. For "the generality of roads" the proportions were as follows:—

Cause of Wear or Injury.	For Fast Coaches. per cent.	For Wagons. per cent.
Atmospheric changes	20	20
Wheels	20	35·5
Horses' feet that draw the vehicles	60	44·5
	100	100

Mr. James Macadam, in 1831, considered that a vehicle of any description, required to carry a great load, say, 5 or 8 tons, ought to have wheels with tires at least 4½ inches wide, flat or cylindrical. He preferred that width to any other width, being of opinion that it was the maximum width of tire that could at the same instant touch the surface of a well-formed road.* For a gross weight of 8 tons, or 2 tons per wheel, it is seen that the weight amounted to 9 cwt. per inch of width—considerably in excess of the working limit assigned by Sir John Macneil.

M. Dupuit tested the width of surface in active contact with tires of various widths, on roads slightly wetted; and he found that the marks of contact with the road traced on the middle of the tires were practically of equal width for various widths of tire, amounting to about 3¼ inches on a tire 6¾ inches wide. The pressure appeared not to have been uniformly distributed over the width of surface in contact; and near the edges of the tires the pressure was nearly nothing. On the tires of a stage-coach, 5½ inches wide, the width of contact was limited to 2½ inches. M. Dupuit ascribes the limitation of width of contact to the natural rounding of the tires by wear and stress, towards the edges, by which the circumference of the wooden fellies themselves are likewise rounded; and he condemns, as illusory, the use of wide tires for the better distributing of excessive loads. He maintains that wide tires are only

* Report of the Select Committee on Steam Carriages, 1831, page 91.

useful when they take a bearing for the whole width, which can only happen when the road is soft, loose, or in bad condition; or when the ruts are filled up and require to be rolled. As a matter of fact, a 6¾-inch tire practically becomes, after a few days, a 5½-inch tire; and, after a few months of wear, a 4¼-inch tire. He adds that 6¾-inch tires wear out after running 10,000 miles, losing 530 pounds of iron; showing that for each pound of iron worn off, 19 miles were traversed:—less than a fourth of the mileage per pound deduced from Sir John Macneil's data for wagons, and countenancing, certainly, M. Dupuit's opinion that great width of tire is illusory, and adds weight without efficiency.

From these and other considerations, it is obvious that the wear of macadam roads must necessarily be much greater than that of paved roads or streets. Mr. Mitchell's remarkable analysis of the material of a macadam road, about to be noticed, places this conclusion in a clear light.

A cubic yard of broken stone metal, of an ordinary size—2 inches or 2¼ inches cube—when screened and beaten down in regular layers 6 inches thick, contains, according to Mr. Mitchell, 11 cubic feet of interspaces, as tested by filling up the metal with a liquid. Herr E. Bokeberg, of Hanover, who, says Mr. Paget, made many very careful experiments on the proportion of vacuity to solid material, found that, in a loosely-heaped cubic yard of broken stones, void space amounted to one-half of the total volume. As the stones became rounded at the corners by wear, the vacuities were reduced to 37 per cent. of the gross bulk, or to 10 cubic feet These results indicate that the interspaces of the new stones were reduced by compression to nearly the same volume as was attained after the corners had been rounded without compression. The operation of consolidation necessarily reduces the thickness of the coating. A large portion of the vacant space becomes filled with mud, which forms the cementing matter, ground from the metal. The original mass of 16 cubic feet of broken stone is, in fact, crushed into every variety of form down to the finest sand,—a small proportion only of the stones remaining of their original dimensions.

Mr. Mitchell* gives the results of an analysis of a portion of the crust—2¼ cubic feet—of the macadamised road in the Mall, in St. James's Park, which was taken up for examination. The component parts of the sample were carefully separated and classed, when it was found that one cubic yard contained—

	Cubic feet.		Per cent.
Mud	11·00	or	41
Sand, with pebbles, not exceeding 3-16 inch thick	2·40	,,	9
Stones, from 3-16 inch to ½ inch . . .	6·56	,,	24
Stones, from ½ inch to 1 inch	4·48	,,	16½
Stones, from 1 inch to 2¼ inch . . .	2·56	,,	9½
Total volume, 1 cubic yard, or	27·00	,,	100

From this analysis, it appears that only 9½ per cent. —say, one-tenth—of the original stone escaped unground; whilst 40 per cent. of it was reduced to mud. These proportions, taken as they stand, are too favourable for the duration of the stone in that instance, for, no doubt, the sample was a sample of the remains of stone, much of which must have been swept or washed off out of sight. Mr. Burt, therefore, cannot be much amiss in his estimate that one-third of the loose road-material used in London is literally wasted by being ground up under the traffic before the actual consolidation of the surface is effected.

The logical inference is that a macadamised road is not properly fit for traffic unless it is condensed, consolidated, and reduced to a hard and regular surface by suitable

* "A New Mode of Constructing the Surface of the Streets and Thoroughfares of London and other Great Cities," 1870.

appliances. The road-roller has long been successfully employed in France and Germany for this purpose. Mr. Paget* states that a well-rolled road-covering contains at least from 70 to 80 per cent. of mass of stone, leaving only from 20 to 30 per cent. of intermediate space, most of which is filled, especially at the top, with clean sand. In a cubic yard, the spaces would amount to from $5\frac{1}{4}$ to 8 cubic feet, which would prove that the spaces are reduced by rolling to nearly a half of the amount when the new metal is not rolled. It is scarcely necessary to insist on the increase of durability, and the clear gain in economy of maintenance, of the road, by efficient rolling. The reduction of gross bulk by the process of rolling, is measured by the difference of the volumes of interspace, $13\frac{1}{2}$ cubic feet, or half a cubic yard, and, say, $6\frac{3}{4}$ cubic feet, the mean of $5\frac{1}{4}$ and 8 cubic feet; that is, the reduction is $(13\frac{1}{2} - 6\frac{3}{4} =)$ $6\frac{3}{4}$ cubic feet, which is one-fourth of the primitive bulk, or one-half of the primitive volume of the interspaces.

It was said by Macadam that the annual wear of metalled roads amounted to from 1 inch to 4 inches of depth. The rate of vertical wear is readily calculated when the area of surface covered and the quantity of metal deposited are known.

Since a cubic yard of loose broken stone contains only one-half of its volume, or $13\frac{1}{2}$ cubic feet, of solid stone, its weight, allowing 12 cubic feet of solid granite to 1 ton, is approximately, $1 \times \dfrac{13 \cdot 5}{12} = 1\frac{1}{8}$ ton.

Again, one cubic yard is equivalent to 36 square yards 1 inch deep; and 1 ton of metal laid without compression to a depth of 1 inch covers an area of $(36 \times \dfrac{1}{1\frac{1}{8}} =)$ 32 square yards. When the metal is laid and rolled, the

---

* "Report on Steam Road-Rolling," p. 10.

primitive volume is reduced, as was found, by one-fourth, and 1 ton of rolled metal laid to a depth of 1 inch covers an area one-fourth less than 32, or (32 × ¼ =) 24 square yards.

On these data, the following rules are framed:—

RULE 1.—*To find the average vertical depth of wear per year of a macadamised road, when the area of surface, and the weight of metal laid per year, are given.*—Multiply the quantity of broken granite laid per year, in tons, by 32 when not rolled; or by 24 when rolled; and divide the product by the area covered, in square yards. The quotient is the average vertical depth of wear per year.

For example, Mr. Redman stated, in 1854, that on the Commercial Road, in the east end of London, 10 yards wide and 2 miles long, the quantity of coating of broken granite required was from 1 inch to 1½ inches deep per year, or a total of 1,200 tons of metal. The heavy traffic of the road was then, as it is now, conveyed on two lines of granite tramway, consisting of four tracks 16 inches in width, occupying together a width of 64 inches, or 1·78 yards. The net width for macadam was, therefore, (10 — 1·78 =) 8·22 yards. The length was (2 × 1,760 =) 3,520 yards; and the area of macadam was (3,520 × 8·22 =) 28,934 square yards. By the rule, for unrolled metal, the average vertical depth of material supplied, and of course worn off, was $\frac{1,200 \times 32}{28,934}$, = 1·33 inches; which is nearly the mean of 1 and 1½ inches given by Mr. Redman. The converse rule follows directly:—

RULE 2.—*To find the area of surface that can be covered by one ton of broken granite, when the thickness of the layer is given.*—Divide 32 by the thickness of the layer, in inches, unrolled; or, divide 24 by the thickness of the layer, in inches, when rolled. The quotient is the area in square yards.

When the quantity is given in bulk, the rule is as follows:—

RULE 3.—*To find the area of surface that can be covered by one cubic yard of broken granite, when the thickness of the layer is given.*—When the metal is not rolled, divide 36 by the thickness in inches; the quotient is the number of square yards that can be covered. When the metal is rolled, divide 27 by the final thickness in inches, to give the required quotient.

The carriage-way of old Westminster Bridge, 30 feet wide, was worn away, according to Mr. Browse, at the rate of 5¼ inches annually; the carriage-way of Bridge Street, adjoining the bridge, wears off 7 inches per year.

The quantity of broken metal worn away annually from the surface of roads, is an exceedingly variable quantity, dependent not only upon the nature of the traffic, but also upon the mode of maintaining the road. A simple method of approximating to the annual rate of wear, when direct evidence is wanting, is suggested by data given by Mr. Paget.* He takes the average price of the five-ounce Guernsey granite metalling, used in London in 1869-70, at 16s. per ton delivered on the roads, and he shows that the cost of the material formed a large proportion of the total cost for maintenance. For instance:—

	Total cost for maintenance.	Cost of material.			
Bond Street	£245	£216	or	88	per cent.
Piccadilly	1384	1359	,,	98	,,
Bridge Street, Westminster	341	305	,,	89	,,
Parliament Street	528	473	,,	90	,,
Victoria Street	869	744	,,	86	,,
Kensington Road	1588	1414	,,	90	,,
Totals	4955	4511	,,	91	,,

showing that about nine-tenths of the whole cost is

* "Report on Steam Road-Rolling," page 32.

expended on the purchase of metal. When the total cost is known, therefore, together with the superficial area of carriage-way maintained, the cost may be converted into tons by dividing it by 16s., and the quantity and depth per square yard, approximately. may be deduced.

CHAPTER IV.

MACADAMISED ROADS:—COST.

LONDON.

THE price of Guernsey metalling, delivered in London, is now (1877) about 17*s.* 6*d.* per ton. An ordinary macadamised road, constructed in the best manner, with metal laid 9 inches deep, costs, at London prices, 6*s.* 3*d.* per square yard.

It appears, from a table elaborated by Mr. Paget,* that in 1869 there were 1,127 miles of macadamised roads in the metropolis, maintained by 39 parishes and district boards of works, comprising an area of upwards of 21¼ millions of square yards. The expenditure for maintenance, exclusive of cost for cleansing and watering, averaged from fourteen of the returns, amounted to £250 per mile of road per year, equivalent to 2*s.* 10*d.* per lineal yard. If the average width of roadway be taken at 10 yards, the cost would be at the rate of 3·4 pence per square yard. The ascertained costs of maintenance appear to have varied from 1¾ pence to 2*s.* per square yard per year for broken granite covering; and for granite, and flint and gravel coverings, taken together, from 1¾ pence to 14⅜ pence. The following table, No. 13, is abstracted from the larger table of Mr. Paget:—

* " Report on Steam Road-Rolling."

TABLE No. 13.—MACADAMISED ROADS IN LONDON: ANNUAL COST OF MAINTENANCE OF SOME OF THESE ROADS IN 1868–69. (Exclusive of Cleansing and Watering.)

Parish or District Board of Works.	Laid with broken granite.			Laid with broken flints or gravel.			Total of macadamised road.		
	Length.	Average width.	Annual cost of maintenance per sq. yd.	Length.	Average width.	Annual cost of maintenance per sq. yd.	Length.	Average width.	Annual cost of maintenance per sq. yd.
	mls. yds	feet.	d.	m. y.	feet.	d.	m. y.	feet.	d.
St. Mary, Lambeth	15 0		5½						
St. George, Hanover Square . .	14 0	40	6½	21 0	40	1·31			
St. Leonard, Shoreditch									
St. Mary, Newington	5 0		13⅔*						
St. Giles, Camberwell	7 0			43 0			50 0		4⅔*
St. James and St. John, Clerkenwell	2 0			9 0					5
St. George the Martyr, Southwark .	15 0	33	2						
St. Martin's-in-the-Fields	1 413	30	10 †						[illegible]
Mile End Old Town							21 0		3½*
St. John, Hampstead							33 0		3½
Westminster District									14⅞‡
Wandsworth—Battersea									2¾
Wandsworth–Tooting, Streatham .									3½
Hackney District .	7 0		8¼*						
Strand District .	0 170		24						
St. Saviour's . .	8 0		1⅔						
Plumstead . . .	27 0	20		13 0			40 0		1¾

* Including Cleansing and Watering.

† Principal thoroughfare of St. Martin's-in-the-Fields cost 2s. per square yard per year.

‡ This cost, of 14⅞d., is the cost for maintenance of Parliament Street, Whitehall, Great George Street, Broad Sanctuary, Victoria Street, Bridge Street, Great Smith Street, James Street, Kensington Road, Exhibition Road, Princes Street, and Rochester Row. Total area—72,329 square yards.

The actual extent of the variation of the cost of maintenance of macadamised roads in London, is not revealed by these tabulated statements. The costs for individual roads or streets,—principal thoroughfares,—exclusive of cleansing, have been stated, by various competent authorities, to be as follows :—

	Year.	Annual cost. Per sq. yd.
Parliament Street	1856	2s. 4d.
Do.	1869	3s. 3d.
Bridge Street, Westminster	1856, 1869	3s. 6½d.
Great George Street, Westminster	1856	6½d.
Westminster Bridge (Old)	1854	2s. 0d.
Piccadilly	1834—63	2s. 5d.
Do.	1870	3s. 6d. to 4s.
Regent Street	1876	3s. 7d.
Cranbourne Street and North Side of Leicester Square	1870	2s. 0d.
Average annual cost		2s. 7¼d.

If the average amount of vertical wear be estimated from these rates of cost, in the manner before indicated, taking 90 per cent. as the cost for metal, the vertical wear would be found to vary from 1 inch to 7 inches per year. But, it must be admitted, this mode of estimation, though suggestive, is inexact.

Suburban Highways.—The Author is indebted to Mr. George Pinchbeck for statistics of the cost for maintaining the suburban highways of the Metropolis, Middlesex side, for one year, in or about the year 1855. The highways were comprised in the following Metropolitan Trusts :— Kensington, Uxbridge, Harrow, Kilburn, Highgate, Hampstead and Islington, Stamford Hill, and Hackney Lea Bridge.

Total length of roads for this account, 123½ miles.

	Total.	Per mile.
Day labour, including the pumping of water	£9,500	£77·03
Digging gravel and preparing materials, per contract	1,830	14·84
Tram labour, including watering	13,500	109·50
Flints, per contract	1,951	15·82
Gravel do.	2,460	19·94
Granite and hard stone	29,430	238·60
Wheelwright work, per contract	640	5·19
Implements	255	2·07
Smiths	158	1·28
Carpenters	251	2·04
Bricklayers	328	2·66
Stand-posts, water-plugs, pumps:—Repairs	365	2·96
Pavior's bill	165	1·34
Water for Roads, from Companies	2,000	16·22
Material and Labour	£62,833 or	£509·5
Rent of Premises	270	2·19
Salaries of Eight Surveyors, at £170	1,360	11·03
Do. Engineer-in-Chief	700	5·68
Do. Solicitor	400	3·25
Do. Accountant	350	2·85
General Expenses	£3,080 or	£25·0
Total Cost	£65,913 or	£534·5

If the width be taken at an average of 30 feet, or 10 yards, the area maintained per mile would be 17,600 yards; and the total cost would amount to 7·29 pence, say 7¼d., per square yard for the year. There were required, in addition, the services of 42 Commissioners, which were rendered gratuitously.

If the width of the roadways be taken at an average of 30 feet, or 10 yards, the area maintained would amount to 17,600 square yards per mile; and the cost per square yard would be as follows:—

	Per square yard.
Material and Labour	6·95 pence
General Expenses	·34 ,,
	7·29 pence.

The cost for material—that is, gravel, flints, and granite and other hard stone—amounted to £35,671 per year, or £289·20 per mile, or about 4 pence per square yard; being 54 per cent. of the total expenses.

For the purpose of forming an estimate of the quantities of granite and flints consumed by wear, Mr. Pinchbeck calculates them from the total costs for the materials, allowing 15 shillings per cubic yard for granite and hard stone, and 4 shillings for flints. He thence deduces that the total quantities consumed were 4,000 cubic yards of granite laid over 100 miles of road, and 10,000 cubic yards of flint laid over, say, 25 miles of road; being at the rate of about 7½ inches deep of granite and of flints for the year.

Local Roads.—The minor roads of the parish of Islington, 36 miles in length, were maintained by the authorities themselves. The average cost during four years was £8,424 per year, or £234 per mile; which, taking an average width of 30 feet, would be at the rate of 3·13 pence per square yard.

BIRMINGHAM.

According to the results of the experience of macadamised roads in Birmingham, in 1853,* it was found that the vertical wear in Bull Street was less than 6 inches per year, where 2,484 vehicles, comprising very heavy traffic, passed in ten hours in one day. With this exception, the greatest amount of wear was 4 inches per year, and, taking the average wear for the whole borough at 2 inches per year, which was a very high rate, the total cost for maintenance was estimated at 4$d.$ per square foot per year. Adding 2$d.$ for the cost of watering and cleansing,

---

* "On Macadamised Roads for the Streets of Towns," by Mr. J. P. Smith, in the *Proceedings of the Institution of Civil Engineers*, vol. xiii. page 221.

the total annual cost was estimated at 6d. per square yard.

But the actual cost for the year 1853, when the actual average wear was 1⅛ inch per year, amounted to a little over 3½d. per square yard. The items of cost are given as follows:—

	Length. Miles.	Area. Sq. yds.	Tons.	Material used.		Cost of material.
Urban streets	34	859,040	20,112	rag stone, &c.,	at 5s. 11d.	£5,949 18 6
Suburban streets and roads	116	1,429,120	36,107	pebbles and gravel,	at 2s. 6d.	4,513 7 6
	150	1,788,160	56,219			10,463 6 0

Mr. Smith takes the volume of materials at the rate of a cubic yard per ton, whence the calculated rates of vertical wear:—

	Material. Cub. yds.	Vertical wear per yd. inches.	Cost of material per sq. yd. per year.
Urban streets	20,112	2	4d.
Suburban streets and roads	36,107	0·91	¾d.
Total	56,219	1⅛	1·40

The cost for labour, haulage, management, &c., inclusive of cleansing and watering, amounted to £16,131 5s. 6d.; and the items of cost may be presented concisely thus:—

Birmingham.	1853.			Per sq. yd.	Per lineal mile.
Macadam	£10,463	6	0	1·40d.	£70
Labour, management, &c.	16,131	5	6	2·16d.	108
	26,594	11	6	3·56d.	£178

showing a total cost of 3·56 pence per square yard per year, as before stated; or £178 per lineal mile of street and road.

The statement in detail, supplied by Mr. Smith, shows the respective amounts for the various items of cost, as here given. The proportional values of these amounts are added as percentages of the total cost:—

	Birmingham.	1853.	Per cent.
1.	Staff of officers and men	£6,979 10 7	26·3
2.	Keep of 63 horses, shoeing, and veterinary expenses	3,276 0 0	12·3
3.	Ragstone and Hartshill stone	5,949 18 6	22·4
4.	Pebbles and gravel	4,513 7 6	17·0
5.	Water	771 3 8	2·9
6.	Cleansing machines and brooms	965 6 8	3·6
9.	Team hire	631 7 2	2·4
10.	Rent	12 0 0	0·0005
11.	Gearing	114 6 4	0·4
12.	Wheelwright	423 7 3	1·6
18.	Stationery	117 7 9	0·4
22.	Sundry accounts	2840 6 1	10·7
	Total cost for maintenance, cleansing, and watering	£26,594 11 6	100·0

Here it is shown that the cost for macadamising materials for maintenance amounted to about 40 per cent. of the total cost. The cost of coarse-grained schist, with 23 per cent. of lime, was 4*s.* 9*d.* per cubic yard. For heavy thoroughfares, it was broken up into cubes of about 2 inches; but, in some cases, where the traffic was more than ordinarily severe, the pieces were 4 inches in thickness. Generally, for streets with less traffic, the thickness was from 1 inch to 1½ inches. The size of the stones was proportioned to the wear and tear of the streets, and was suited to the season of the year.

In the management of repairs by Mr. Smith, they were executed during the winter months—November, December, January, and February—when the roads were in their wettest condition; and during the other months of the year, the horses were employed in bringing, and the men in preparing, the broken stone and gravel. The new material, laid in the wet season, bonded, and became more rapidly consolidated, than if it had been laid down during drier weather, and there was less need for watering it. The consequence was that, during summer, loose stones were scarcely to be seen upon the roads.

Whilst the average cost for maintaining the whole of the macadamised streets and roads of Birmingham has been shown to be about 3¼d. per square yard per year, the following estimates were framed by Mr. W. Taylor, of the first cost, and the cost for maintenance and cleansing the streets of the greatest traffic in Birmingham, when macadamised, and when paved with granite sets; taking a period of fifteen years for the duration of the sets, and taking a vertical wear of 4 inches of macadam annually:—

Birmingham.	Macadam Per square yard.		Per year d.
Macadam 6 inches thick, first cost	1s. 6d.	or	1·2
Coating laid twice a year, 4 inches thick, annually, 14 years at 1s.	14s. 0d.	„	11·2
	15s. 6d.	„	12·4
Cleansing, 15 years at 4d.	5s. 0d.	„	4·0
Total cost for macadam, in 15 years	20s. 6d.	„	16·4

	Paving. Per square yard.		Per year. d.
Paving, first cost	7s. 6d.	or	6·0
Repair, 15 years	1s. 6d.	„	1·2
	9s. 0d.	„	7·2
Deduct, value of old stone	2s. 6d.	„	2·0
	6s. 6d.	„	5·2
Cleansing, 15 years at 1d.	1s. 3d.	„	1·0
Total cost for paving, in 15 years	7s. 9d.	„	6·2

Derby.

The streets of Derby are, or at least they have been until recently, macadamised, with one exception, which is paved. The following instructive table gives particulars of cost for maintenance and cleansing of twelve principal streets in Derby, and the proportional traffic.*

* Slightly adapted from a Table given in a Report of Mr. E. B. Ellice-Clark, Surveyor to the Borough of Derby, May, 1876.

TABLE No. 14.—DERBY.—MACADAMISED STREETS, COSTS FOR MAINTENANCE AND CLEANSING.

Name of Street.	Area.	Material.	Annual cost for maintenance per square yard.	Annual cost for cleansing per square yard.	Total annual cost for maintenance and cleansing per sq. yd.	Number of vehicles in 12 hours.
	sq. yd.		s. d.	d.	s. d.	vehicles
Bridge Gate .	2931	Broken boulders	1 6	6½	2 ½	1268
St. Helen's Street	1203	Do. granite .	1 2½	5¼	1 7¾	585
Wardwick . .	984	Do. do.	1 2½	8½	1 11	1258
Iron Gate . .	1562	Do. do.	7¼	4	11¼	1506
Corn Market .	1695	Do. do.	7¼	3½	10¾	2031
St. Peter's Street	3230	Do. do.	1 2½	3	1 5½	2762
London Road .	4859	Do. boulders and granite	7¼	2	9¼	1272
London Street .	2847	Do. do.	10¾	3	1 1¾	1200
Tenant Street .	1245	Do. do.	1 2½	3¾	1 6¼	1250
St. James's Street	995	Do. do.	5½	4½	10	711
Siddals Road .	4728	Do. do.	7¼	3	10¼	1931
Averages for macadam	—	———	11	4¼	1 3¼	1431
Derwent Street, paving down 7 years	3483	Sets 3" × 5"	nil	2¼	2¼	1658

It is to be remarked, in this table, that the boulder macadam of Bridge Gate incurs a higher cost for maintenance than any of the granites. Wardwick, upon which the greatest cost for cleansing is expended, is cleansed daily. The average annual cost for maintenance for eleven macadamised streets is 11*d.* per square yard, and for cleansing 4¼*d.*; whilst, for the seven years during which the paving has been down, it has not cost anything for maintenance, but it has cost for cleansing 2¼*d.*

Mr. Ellice-Clark has formed estimates of the cost for construction and maintenance of granite-set paving, in substitution for macadam, for five streets in Derby; from which the following table, No. 15, is compiled:—

TABLE No. 15.—DERBY:—ESTIMATED COST FOR PAVING AND MAINTAINING FIVE STREETS.

Granite Sets 3 inches wide by 5 inches deep.

Name of Street.	Duration of Pavement.	First cost per square yard.		Cost for maintenance per square yard.		Total cost per square yard.		Cost for cleansing per square yard per year.	Total cost for construction, maintenance, and cleansing per square yard per year.
		Total.	Per year.	Total.	Per year.	Total.	Per year.		
		s. d.	d.	s. d.	d.	s. d.	d.	d.	s. d.
Bridge Gate	24	13 0	6·5	7 6	3·8	20 6	10¼	2¼	1 0½
St. Helen's Street	29	13 0	5·4	7 6	3·1	20 6	8½	2	10½
London Road	32	12 6	4·7	6 6	2·44	19 0	7⅜	1	8⅜
Tenant Street	28	13 0	5·5	7 2	3·1	20 2	8½	1½	10
Siddals Road	32	12 6	4·7	6 6	2·44	19 0	7⅜	1¼	8⅜
Averages	29	—	5·36	—	3·0	—	8·36	1·6	10

The first cost includes the cost for a foundation of concrete. No deduction has been made by Mr. Ellice-Clark for the value of the old materials; this is taken as equivalent to the cost of removing the existing macadam. In the estimates for maintenance, provision is made for taking up, re-dressing, and re-laying the stones once during the period.

From the foregoing data, a comparison is readily framed between the costs in perpetuity for macadam already laid and new granite-paving, including a foundation of concrete, for the five streets named in the last table:—

TABLE No. 16.—DERBY:—COMPARATIVE COSTS FOR GRANITE PAVEMENT AND MACADAM.

Pavement of Granite Sets, 3 inches by 5 inches deep.

Name of Street.	Duration of Pavement.	Total cost for pavement per sq. yd., including foundation of concrete.		Total cost for macadam, already laid, per square yard.		Number of vehicles in 12 hours.
		For the whole period.	Per year.	For the whole period.	Per year.	
	Years.	s. d.	s. d.	s. d.	s. d.	Vehicles.
Bridge Gate	24	25 0	1 ¼	49 0	2 ½	1268
St. Helen's Street	29	25 4	10½	47 9	1 7¼	585
London Road	32	21 8	8⅛	24 8	9¼	1272
Tenant Street	28	23 8	10	42 7	1 6¼	1250
Siddals Road	32	22 4	8¾	18 6	10¼	1931
Averages	29	24 2	10	39 11	1 4½	1261

SUNDERLAND.

Mr. D. Balfour[*] reported, in 1876, on the cost for maintenance of the highways of the Sunderland and Houghton-le-Spring District, before and after the passing of the Highway and Turnpike Act of 1869. For the maintenance of 77½ miles of highway, the following were the respective costs. The last two lines are added by the writer:—

AVERAGE ANNUAL EXPENDITURE.

	During last 8 years, Old Act.	During last 6½ years, New Act.	During last 8 years, New Act.
Materials and Labour	£1,903	£1,735	£1,509
General Charges	—	284	277
Total Cost	£1,903	£2,019	£1,786
Do. do. per mile	£24·5	£26	£23
Do. do. per square yard allowing width of 30 feet	3·34d.	3·55d.	3·14d.

[*] "Road Legislation and Surveying." By D. Balfour, A.I.C.E., Surveyor and Engineer to the Sunderland and Houghton-le-Spring District Highway Board, 1876.

Districts near Edinburgh, Glasgow, and Carlisle.

Mr. James H. Cunningham, in his paper already noted, gives the following particulars of cost of construction of macadam roads in the south of Scotland :—

Country Roads (1861—1874) cost from 1s. 2d. to 1s. 6d. per square yard.

Town Roads, founded on shivers, or on hard set stone :—

    15 inches of metal (1871) cost 1s. 9d. per square yard.
    16   ,,   ,,   (1874) ,, 2s. 4d.   ,,   ,,
    18   ,,   ,,   (1873) ,, 2s. 11½d. ,,   ,,

Of whin metal, from 2 inches to 2½ inches thick,—the average price (1861—1874) was 4s. 10d. per cubic yard. In Edinbugh (1874) the price was 5s. 9d. to 7s. 3d. per ton, averaging 6s. 8d. per ton, and as, according to Mr. Cunningham, one cubic yard of whin metal, containing 40 per cent. of vacant space, weighs 27 cwt., the average price is at the rate of 4s. 10d. per cubic yard,—the same as the average already given.

CHAPTER V.

CONCRETE ROADS.

Mr. Joseph Mitchell designed a concrete macadam surface, in which the spaces, otherwise vacant, and ultimately filled with the muddy cementing matter of worn macadam, are, in the construction of the concrete road, filled with an admixture of Portland cement, or other hydraulic cement grout. The concrete thus formed rapidly becomes a uniform and impervious mass, which is wholly unaffected by heat or moisture. It is mixed in these proportions:—

Broken stones	4 measures.
Clean sharp sand	$1\frac{1}{4}$ to $1\frac{1}{3}$,,
Portland cement	1 ,,

So, for a cubic yard, or 27 cubic feet, of broken metal, $6\frac{3}{4}$ cubic feet, or $1\frac{1}{2}$ barrels (of $4\frac{1}{2}$ cubic feet), of Portland cement, are required. The broken metal should be of the hardest quality, of uniform size, thoroughly screened; and it should, when in the screen, be dipped up and down in a large tub of water, and then thrown on the platform on which the concrete is to be made. Cement of the best quality must be employed, and the sand should be sharp, clean, and gritty. The surface of the ground is brought to form, and rolled several times. The concrete is then laid on the surface in a layer 3 or 4 inches, and is left for three days to harden. The second layer of 3 or 4 inches is next laid on the first, and immediately rolled to form with a heavy iron roller, as heavy as two or three men can draw. The cement should be left for three weeks, to allow it to become quite hard, before the road is opened for traffic, although a week has been found to be a sufficient interval.

Mr. Mitchell states that a concrete road, 7 inches deep at the middle and 5 inches at the sides, is sufficient for ordinary traffic. For heavy traffic, a depth of 8 inches is recommended.

The first piece of concrete road was laid, in 1865, in Inverness, and consisted of 45 lineal yards of the approach to the goods station of the railway. In 1870, after the road had been under traffic for $4\frac{1}{2}$ years, it was reported that the wear of the surface was scarcely appreciable, whilst the adjoining macadamised road had been coated frequently every year.

Another specimen, 50 yards long and 15 yards wide, was laid in 1866, on George IV. Bridge, Edinburgh, where the traffic is heavy and continuous. At the end of $3\frac{1}{2}$ years, under traffic, the surface was perfectly sound and immovable.

The amount of vertical wear, during the periods above-named, appears not to have exceeded $\frac{1}{4}$ inch. But Mr. J. H. Cunningham, writing in January, 1875,[*] stated that it was then much worn at the surface, in consequence, he thought, of its great hardness and rigidity.

A concrete road, of 6 inches average depth, would cost in London, by Mr. Mitchell's estimate, 6s. 10d. per square yard, against 6s. 3d. per square yard for an ordinary macadamised road, constructed in the best manner, with 9 inches of metal.

[*] See Mr. Cunningham's paper on "Streets," in the *Proceedings of the Edinburgh and Leith Engineers' Society*, 1874-75, page 18.

CHAPTER VI.

MACADAMISED ROADWAYS IN FRANCE.

M. Dumas, in the article already referred to (page 20), maintained that, if a macadamised road be properly and actively watched, levelled, and cleaned, a foundation, as such, is of no special utility, and may be suppressed; and that the road should be simply a uniform bed of small materials.

The width of the old roads, he says, was excessive, amounting sometimes to nearly 80 feet, with a pavement in the middle, from 16 to 20 feet wide. They are now made of widths of from 8 to 14 metres, or 26 to 46 feet. Type-sections of French and Belgian roads and streets are shown in Figs. 42 to 44.

There is no need, says M. Dumas, for great thickness of

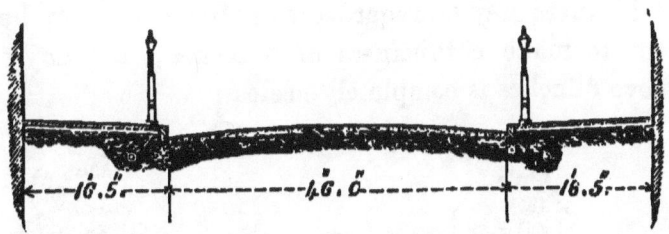

Fig. 42.—Section of Rue de Rivoli, Paris.

road, provided that a compact and impermeable mass be formed, completely shielding the ground from moisture. The least thickness that may be given depends on the size of the materials. If the pieces are 2 inches or 2¼ inches diameter, mixed with pieces of smaller size, a solid roadway may be formed, of a thickness not exceeding 4 inches. Inferior quality of material, bad ground, or heavy traffic, does not imperatively demand greater thickness, if the road be well and continuously maintained. Four inches

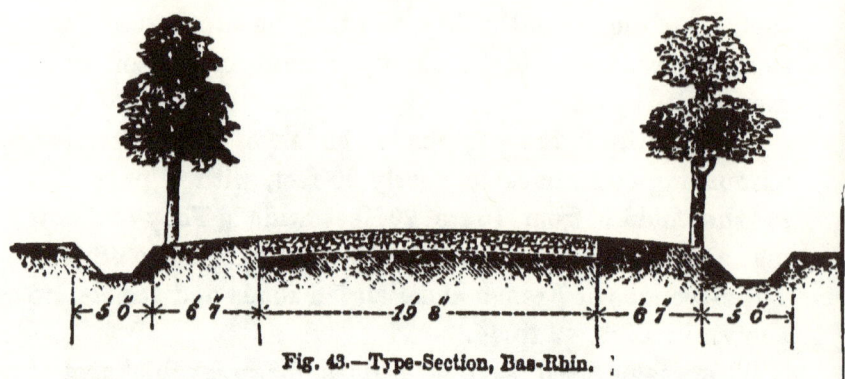
Fig. 43.—Type-Section, Bas-Rhin.

of thickness may be regarded as sufficient: it may be prudent to make a thickness of 6 inches; and any excess above 8 inches is completely useless.

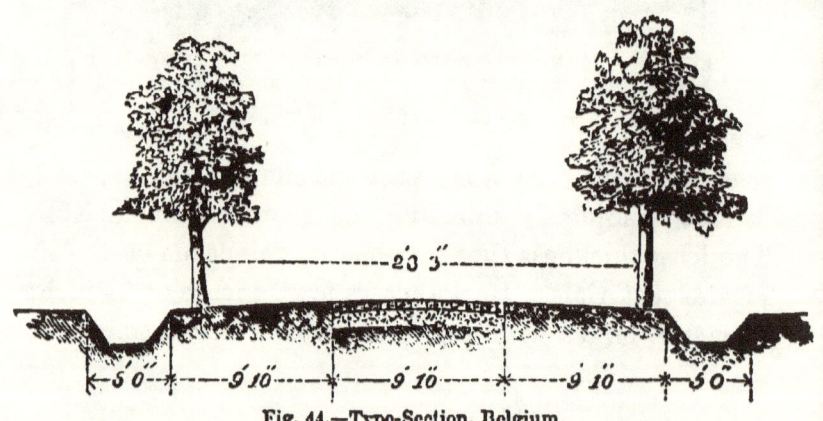
Fig. 44.—Type-Section, Belgium.

The curvature of the old roads was excessive. The rise was from 1 in 14 to 1 in 12. The most convenient rise is 1 in 33, which is much more than sufficient for a well-kept road. The minimum appears to be 1 in 50; which is the slope upon which the weight is in equilibrium with the frictional resistance on roads in good condition. When newly made, the rise may be 1 in 25, to allow for the greater wear at the centre than at the sides. The form of the base, or surface of the ground, may follow the curvature of the surface; but M. Dumas prefers a horizontal line, as in all respects preferable. A road 10 metres, or 33 feet wide, with a rise of 1 in 33, and an average thickness of 4 inches, on a horizontal base, would have, according to M. Dumas, a thickness of 8 inches at the middle or crown, and 2 inches at the sides. But these measurements are not consistent with each other.

The material for the road must have the qualities of hardness, and facility for binding. The largest pieces must pass through a 2½-inch ring; the smaller pieces occupy the interspaces. The 2½-inch gauge is perhaps too large for very hard material, but it is more than sufficient for the softer materials. The smallest *débris* of broken material should be preserved, and spread on the surface. For gravel, the lower limit of size is fixed by the use of a sieve having ¾-inch meshes; otherwise there may be nothing but sand for binding. The smallest pieces should be on the upper surface; if not, the upper pieces are broken. It is well, finally, to break the pieces on the made surface: the *débris* binds.

For the consolidation of the road, it should be rolled and watered. The cylinder of the roller should be from 6 feet to 6¼ feet in diameter, and 5 feet wide; weighing, empty, 3 tons, and full, 6 tons. The maximum weight, when loaded, should be from 8 to 10 tons. These weights give a pressure varying from 112 pounds to 370 pounds per inch of width; the sufficiency of which has been proved by experience. The empty roller is first used; then the full roller; and, lastly, the weighted roller. Sand or other binding, with water, are thrown on the surface at intervals; the water helps to bind the new material. The material binds most speedily when the thickness is 4 or 5 inches. For this thickness 8 or 10 traverses are sufficient:—two traverses dry, with the empty roller; two, with a layer of detritus and the empty roller; two with the full roller, two with the loaded roller, and one at an interval of from 8 to 15 days after the road is opened for traffic. The same number of horses perform the work of rolling from first to last. But the surface of inclined portions may be injured by the feet of the horses; and the injury is partially prevented by increasing the number of horses. It is nearly impossible to roll over inclines steeper than 1 in 20.

If the metal is more than 4 or 5 inches thick, it should be deposited in two successive layers, to be rolled successively; although the first layer should not be completely rolled; and the second layer should be laid after the surface of the first has been roughed by the traffic.

The surface is kept up by the use of a stamper or rammer, weighing from 15 to 20 pounds, which is 8 inches in diameter at the base, and is shod with iron. In some cases, stamping may be applied with advantage, instead of rolling.

A road constructed on the system above described, 4 inches thick, is superior to a road 8 or 10 inches in thickness, consolidated by wheels, and with successive additions of material.

The flanks of the road or the bermes (*accôtements*), consisting of the natural ground, are fit for traffic during the greater part of the year.

"All these proceedings," says M. Dumas, "have, for their basis, the principle of the *maximum of beauty*."

CHAPTER VII.

STONE PAVEMENTS—CITY OF LONDON.

It has been stated that the streets of London were, previously to the introduction of granite sets, paved with pebbles or boulder-stones, which were bedded in sand, and constituted pebble-paving. This pavement was gradually superseded by a pavement of irregular blocks of stone, and ultimately by rectangular granite sets, or "cubed granite," as they are sometimes designated; until, in 1848, there remained only one mile of pebble-paving in a total length of 50 miles, in the City alone. Granite sets of comparatively large dimensions were at first employed. They were from 6 to 8 inches in width on the surface, by from 10 to 20 inches in length, with a depth of 9 inches. As originally laid, they were merely laid in rows on the subsoil, and, after the usual process of grouting and ramming, the street was thrown open for the traffic which was expected to perform the last duty of the pavior, and to settle each stone upon its bed. The large wooden rammer, of 84 pounds weight, was obviously insufficient for the purpose of enabling the pavement to resist, without further movement, the percussion of heavily weighted wheels. In 1850, however, and probably for some time previously, it had become the general practice to make a good substratum of "hard core," consisting of shivers, broken stone, brick rubbish, clinker, or other hard material,—not necessarily of the size of macadam, nor, on the contrary, laid on a pavement on the Telford system of road-making, but a medium between these extremes of practice, such as Telford himself recommended in 1824. The substratum of hard core has been usually laid to a depth of from 9 inches to 12 inches; but a thickness of 15 inches has been laid down in the principal streets of the City. Upon

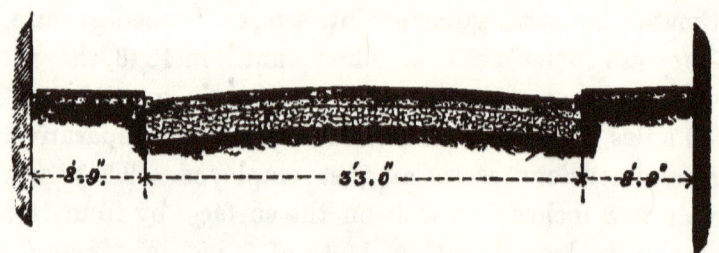

Fig. 45.—Stone Pavement of King William Street, City of London.

the hard core was laid a stratum of sand, into which the stone sets were bedded. Fig. 45 shows a section of King William Street, as originally paved.

The following table, No. 17, gives the dates of the formation of some of the earliest carriage-way granite pavements laid in the leading thoroughfares in the City of London, with their superficial dimensions, and the lengths of time that they remained before being removed to other places.*

* This table is compiled from the Reports of Lieutenant-Colonel Haywood, Engineer and Surveyor to the Commissioners of Sewers of the City of London. The writer is indebted to this valuable series of Reports for the greater proportion of the materials employed in the preparation of what follows with respect to the streets of the City of London.

TABLE No. 17.—CITY OF LONDON:—EARLIEST CARRIAGE-WAY GRANITE PAVEMENTS IN LEADING THOROUGHFARES.

Aberdeen Granite Sets, 6 inches wide by 9 inches deep.

Name of Thoroughfare.	When laid.	When taken up.	Time down.	Remarks.
	Year.	Year.	Years.	
Bridge Street	1828	1818	20	
Ludgate Hill	,,	1814	16	
Ludgate Street	,,	1847	19	
Bishopsgate St. Without	,,	1853	25	
Poultry	1829	1841 & '43	11 & 13	Removed for wood paving
Cheapside	,,	1843 & '44	14 & 15	Do. do. 1843 to 1846-47.
Farringdon Street	1830	—	—	Relaid in 1858
Skinner Street	,,	1845	15	
Newgate Street	,,	1842	12	Removed for wood paving.
Aldersgate Street	1831	1857	26	
St. Paul's Churchyard	,,	1847	16	
Holborn	,,	1849	18	
Threadneedle St.	,,	1862	31	
Fleet Street	1832	1846	14	
Fenchurch Street	,,	1852	20	
Cannon Street Walbrook	,,	1856	24	
Little Moorfields (sets 5 in. by 7)	1833	—	—	Still down in 1853
King William St.	1834	1849	15	
Princes Street	1835	1850	15	
Mark Lane	1837	1856	19	
Houndsditch	,,	1853	16	
Fore Street	1838	1856	18	

Nearly the whole of the stones which formed these early pavements was removed to and laid in places of secondary traffic within the City, where their duration has been as great as, and in some instances greater than, it had already been in the original thoroughfares.

The cost for reparation, up to the year 1840, of 36 streets laid with granite sets 6 inches wide and 9 inches deep, including nearly all the streets named in the preceding table, was calculated by Mr. Kelsey, formerly the City Surveyor, and the results are given in the following table, No. 18.

TABLE No. 18.—COST FOR REPARATION OF CARRIAGE-WAY PAVEMENTS, Laid with 6-inch Granite Sets, in the City of London, up to the year 1840.

Name of Street.	Average cost of reparation per square yard per year.	Name of Street.	Average cost of reparation per square yard per year.
	pence.		pence.
Aldgate High Street	over ¼	Holborn . . .	over 1
Aldgate Street, and Aldgate . .	over ¾	King Street, Snow Hill . . .	over ½
Aldersgate Street .	over ¼	King Street, Cheapside . . .	not ½
Arthur Street West	¾	Leadenhall Street .	nearly 2
Bishopsgate Street Without . .	over ½	Long Lane . .	not ¼
Bishopsgate Street Within . .	over ½	Lombard Street .	nearly 1¼
Budge Row and Watling Street .	1⅛	Ludgate Hill and Street . . .	not ½
Cannon Street . .	nearly 1	Newgate Street .	3
Cheapside and Poultry . . .	1½	New Bridge Street (part of) . .	not ¾
Coleman Street .	not ¾	Old Bailey (part of)	2½
Cornhill	over ½	Pavement, The . .	not ⅛
Dowgate Hill . .	nearly ½	Queen Street . .	not ¾
Farringdon Street .	not ¼	Shoe Lane (part of) .	not ¾
Fenchurch Street .	over ¾	St. Paul's Churchyard . . .	over ¾
Fleet Street . .	nearly 1¾	Threadneedle Street	not 1
Fish Street Hill .	not ¼	Walbrook . .	2¾
Gracechurch Street .	not 1	Wood Street . .	over 1
Holborn Bridge and Skinner Street .	not ½	Average cost for repair, say . .	1 penny
Holborn Hill . .	1¾		

The average cost for repair, according to the table, may be taken at 1*d*. per square yard per year. The first cost was considerably affected by the nature of the pavement laid,—the size of the stones, the nature of the granite, and the market price of the material. Colonel Haywood estimated that, for stones 9 inches in depth, it ranged from 11*s*. to 17*s*. per square yard.

The long continuance of the system of paving with large blocks resulted from the experience of their great durability and economy in first cost. But they did not afford

INTRODUCTION OF THREE-INCH SETS. 173

sufficient foothold for horse-traffic. Granite sets of less width were subsequently laid:—they were 5 inches and 4 inches in width; and, finally, sets of only 3 inches in width were laid. The 3-inch sets, with a depth of 9 inches, although they were considered to be the least durable, and the greatest in first cost, proved to be by far the safest for paving, and they gave more satisfaction than the wider sets. The merit of their introduction is due to Mr. Walker, under whose direction Blackfriars Bridge was, in 1840, paved with 3-inch granite sets. Mr. Walker attached great importance to the obtainment of a solid substratum. Blackfriars Bridge, to the paving of which allusion has already been made, was closed for some weeks, in order that the concrete foundation might have time to set and harden, before the pavement was laid down. The narrow granite sets were laid with a great degree of accuracy, and the whole mass was bedded as if it was composed of bricks: the stones were bedded in good mortar, and the joints were well filled with it, and in consequence of the careful workmanship, with the use of narrow stones, the work remained good for 13 years, at the end of which time it was lifted.

From that time—1840—the use of 3-inch sets was extended, until, in 1848, there were, in the City of London alone, three miles of carriage-way paved with them. The following analysis, prepared by Colonel Haywood, shows the lengths of the various kinds of pavement in the City of London, in November, 1848:—

CITY OF LONDON PAVEMENTS, IN NOVEMBER, 1848.

	Miles.
Carriage-way pavement of pebbles or boulders	about 1
Carriage-way pavement of 6-inch, 5-inch, and 4-inch granite sets (nearly the whole being 6-inch sets)	„ 28½
Carriage-way pavement of 3-inch sets	„ 3
Wood pavement, various kinds	„ ¾
Macadamised road, in Finsbury Circus	„ ¼
Flagstone paving, courts, alleys, &c.	„ 16¼
Total	50

In 1851, there were 51 miles of public ways in the City, containing

441,250 square yards of carriage-way, averaging for, say, 34½ miles, about 22 feet wide.
328,907 square yards of footway.
———
770,157

In 1866, Colonel Haywood reported that there were then 50 miles of public way:—

	Miles.
Main thoroughfares	7
Collateral thoroughfares	28
Minor streets, courts, alleys, passages, &c.	15
	50

Area of carriage-way, about	390,260	square yards.
Do. footway ,,	309,018	,,
	699,278	,,

Regarding the question in all its bearings, Colonel Haywood concluded that the 3-inch granite sets, being safest, as giving the best foothold, were the best for large towns with great traffic. On such a pavement, horses were less strained, there was less wear and tear of vehicles, and a greater degree of quietness and general comfort. They made the most even pavement, and they retained an even surface, longer than any other stone pavement that had been tried.

In adopting, in 1848, the conclusion that 3-inch sets— that is, granite sets 3 inches in thickness, with a depth of 9 inches—were the best for large towns, Colonel Haywood had before him the experience of a trial piece of what was called the Euston pavement in Mount Sorrel granite, which had been laid in 1845, in Watling Street, and relaid in 1848. This pavement derived its name from the circumstance of its having been laid about the year 1843, at the departure side of the Euston Station of the London and North Western Railway. The same sort of pavement had previously been tried experimentally, and for the first time, in 1838, at Birmingham, at the crossing of a street, where heavy loaded wagons were constantly passing over it, and, in 1850, it was stated by Mr. William Taylor[*] to be in as perfect a condition as when it was first laid. Mr. Taylor's motives for advocating the Euston pavement, were based upon the principle advocated by Macadam,—to provide a foundation possessing a certain amount of elasticity, but of sufficient strength to support the surface material; but he substituted "one stratum of solid granite," in the form of paving-stones 4 inches deep, for an equal depth of macadam or broken ring-stone:—thus endeavouring to combine the elasticity, so-called, of a macadam road, with the durability of a paved road. As laid at Euston, the ground was first removed to the depth of 16 inches below the intended level of the pavement, the bottom being formed to the convexity of the intended surface of the street. A layer of coarse gravel, 4 inches thick, was spread upon the bottom; and compressed by being rammed equally throughout. Upon this layer was placed another layer, 4 inches thick, of gravel mixed with a small proportion of chalk, or of hoggin, and rammed likewise. A third layer of the same materials as the second, but of a finer quality, was distributed, and the whole was well rammed together, to form a solid and level surface ready to receive the pavement. Upon a 1-inch bed of sand, stones of Mount Sorrel granite were laid, measuring 3 inches wide, averaging

---

[*] *Proceedings of the Institution of Civil Engineers*, 1849-50; vol. ix., p. 214:—"Observations on the Street Paving of the Metropolis; with an Account of a peculiar system adopted at the London and North-Western Railway Station, Euston Square," by William Taylor.

4 inches in length, and from 3 to 4 inches deep, neatly dressed and squared; and carefully and closely jointed, in order to prevent any single stone from rocking in its bed. A rammer of 55 pounds weight was then applied over the whole surface, after which the pavement was covered with a sprinkling of screened gravel. The maximum cost of the Euston pavement, the foundation included, was said to amount to 12s. per square yard. Mr. Dockray, resident engineer of the railway, in testifying to the excellency of the Euston paving, stated that the carriage-way had previously been substantially executed, according to the old system, with stones 8 inches deep, laid upon a substratum of concrete; but that the stones became much rounded on the upper surface, and were required to be removed. In the Euston pavement he stated that, by reason of the smallness of the stones, they never became rounded on the upper surface by wear, as was the case in ordinary pavements of large stones. The Euston pavement was subsequently laid in the large esplanade in front of the booking offices at Euston station. The carriage-way traffic at Euston station has of course been of a light order,—limited to carriages, cabs, and omnibuses.

Reverting to Colonel Haywood's piece of Euston pavement, in Watling Street, he had experienced a difficulty in obtaining a satisfactory test of the durabilities and cost of various pavings; and, with a view towards the settling of the question of the most useful dimensions of granite sets, for the carriage-ways of the City, he recommended that different sorts of stone-paving should be laid in Moorgate Street. The experimental pieces were laid, all at the same time, in October, 1848. The situation was well chosen, for the different pieces of paving were as nearly as possible subjected to the same amount and kind of traffic, with the usual disturbing and destructive influences. The dimensions and the first costs of the experimental paving are exhibited in Table No. 19.

TABLE No. 19.—CITY OF LONDON :—EXPERIMENTAL PAVING IN MOORGATE STREET, 1848—56.
Dimensions and First Cost.

Name of Granite.	Dimensions of Sets.			First cost per square yard complete.
	Depth.	Width or thickness at the surface.	Length at the surface.	
	inches.	inches.	inches.	s. d.
1. Blue Aberdeen .	9	4	8 to 12	13 8
2. Mount Sorrel .	9	4	8 to 12	14 9
3. Blue Aberdeen .	9	3	8 to 12	16 3
4. Mount Sorrel .	9	3	8 to 12	17 0
5. Blue Aberdeen .	5	3	5	10 9
6. Mount Sorrel .	5	3	5	12 0

NOTE TO TABLE.—The cost of the substratum is not included in the above prices, but the whole of the specimens were laid upon a substratum of equal depth and compactness, and the cost for a foundation may be taken for all practical considerations, as being the same for the whole. The prices given, therefore, are for the new granite surface only, laid and grouted with lime and sand, and left complete.

The pavings were duly and carefully maintained during the eight years from October, 1848, to October, 1856, under Colonel Haywood's personal superintendence. The accounts of charges for maintenance were accurately kept, and the results, with additional particulars, are given in Tables Nos. 20 and 21.

TABLE No. 20.—CITY OF LONDON:—EXPERIMENTAL PAVING IN MOORGATE STREET, 1848—56.
Cost for Reparations.

Name of Granite.	Cost for reparation in 8 years, per sq. yrd. per year.	Annual interest at 5 per cent. on first cost, per sq. yrd.	Cost for reparation and interest together, per sqr. yard per year.	Date of first reparation after laying down the several pavements, in October, 1848.
	d.	d.	d.	
1. Blue Aberdeen	2·076	8·20	10·27	In quarter ending M'mas, 185_
2. Mount Sorrel	·96	8·70	9·66	In half year „ „ 185_
3. Blue Aberdeen	4·62	9·74	14·36	In „ „ „ 185_
4. Mount Sorrel	1 872	10.20	12·07	In „ „ „ 185_
5. Blue Aberdeen	6·684	6·46	13·14	In quarter ending Mar. 25, 185_
6. Mount Sorrel	5·844	7·20	13·04	In half year ending M'mas, 185_
Averages	3·672	8·41	12·08	

TABLE No. 21.—CITY OF LONDON:—EXPERIMENTAL PAVING IN MOORGATE STREET, 1848—56.
Data showing Comparative Values.

Name of Granite.	Number of stones in a square yard.	Area of surface of one stone.	Cubic contents of one stone.	Area relaid in 8 years.	Area of new stones laid in 8 years.
	stones.	square inches.	cubic inches.	per cent.	per cent.
1. Blue Aberdeen	29	44·7	402·2	119 8	1·63
2. Mount Sorrel	25	51·8	466·6	41·8	·75
3. Blue Aberdeen	35	37·0	333·3	243·6	3·93
4. Mount Sorrel	34	38·1	343·1	108·7	·82
5. Blue Aberdeen	51	25·4	127·0	253·2	20·17
6. Mount Sorrel	57	22·7	113·7	308·1	7·07

From the foregoing data, it appears that Aberdeen stone required earlier and more extensive repair and renewal than Mount Sorrel stone; and that the pavings composed of the smallest stones needed more reparation, and the insertion of a greater quantity of new stone, and cost more for repair, than those composed of the larger stones. Colonel Haywood states that these conclusions accord with the results of his general experience; but that the single conclusion which could safely be deduced was that, within the limits of the sizes of stones laid in the experimental paving, the cost for repair upon similar paving was inversely as the size of the stones.

In 1855, the experimental paving was to a great extent relaid; and in 1858 it was wholly relaid. In 1864, when it had been down altogether 16 years, it was removed. The first cost averaged 14*s*. 3*d*. per square yard, and the total cost for laying and repair during the 16 years it was down, not including any charge for interest on first outlay, was—

First Cost . .	10¾d.	per square yard per year.
Cost for Repair . .	4½d.	„ „
Total Cost . .	15¼d.	„ „

Referring to Table No. 20, it is seen that the average annual cost for repair, during the first 8 years, was only 3·67 pence per square yard. The greater cost during the last period of 8 years, arose no doubt from the inferior condition of the pavement, due to the loss of depth of the sets by wear; and to some extent, also, from the augmented traffic.

Granites of various qualities have been tried for carriage-way pavements in the City:—Aberdeen, Guernsey, Herm, Devonshire, Cornish, Mount Sorrel. The harder and more durable granites, like the Guernsey and the Mount Sorrel granites, though the more economical, possess the fault of slipperiness when set in pavement; the less durable granites wear roughly, and therefore afford a better foothold for horses. Hence it is that, for the sake of public convenience, the hardest and most durable granites are not employed. The hardest granites have invariably caused so much dissatisfaction that they have had to be removed before they had been down many years. For instance, Penmaenmaur greywacké paving was laid in Newgate Street in 1851, and it was removed after a trial of 2¼ years, as it was the cause of repeated complaints of slipperiness and noisiness. The Aberdeen blue granite sets have for the most part been employed in the construction of City pavements; they are considered to be the best, taking together the first cost, the durability, and the absence of slipperiness.

Since the cost of a pavement depends upon the material of which it is formed—the width of the street, the extent and nature of the traffic, and other conditions—it follows that, in no two streets is the endurance or the cost the same, and the difference between the highest and the lowest periods of endurance, and amount of cost, is very considerable. The practice pursued, almost uniformly, with respect to the rotation of granite paving, is to lay the new granite in the main thoroughfares; when their pavements are considerably worn, and the stones require to be reworked before they can be advantageously relaid, or when an entire relay of the surface is needful, the worn granite is removed, and new granite laid in its place. In this way, it most commonly occurs that the pavements in main thoroughfares are removed before they are worn out. The old material is taken to the stone-yard, where it is mixed with the general stock, reworked, sorted into sizes, and laid in other, and secondary, thoroughfares when needed. Thus the duration or life of the stones may be doubled, or more than doubled. "Indeed, with the exception of the portion worn off by the friction of the traffic, not a fragment of granite paving may be said to be lost. After passing its first years in a leading thoroughfare, it goes into a secondary thoroughfare until completely worn down and rounded, and will even then command a price of from 1s. 6d. to 2s. 6d. per square yard. Not even a fragment that is knocked off the component stones, when undergoing the operation of being dressed into shape, is lost; as it is made available either directly for macadamising, or for forming substrata to other pavements; or, if such employment cannot be found for it, it will always command a good price by its sale. In truth, granite can only be said to be worn out when it has been broken up for macadamisation, and then crushed into powder by the vehicles."

It is due to the system of rotation, above described, which largely conduces to the general convenience, that although the general cost for repairs is accurately known, the cost of most of the pavements individually, during a term of years, cannot be arrived at, excepting by estimation.

Moreover, it is to be borne in mind, on the question of the durability and cost of stone pavements, that the traffic of the City has not been stationary, but that it has gradually increased in the course of years, and that, to arrive at a just conclusion, the traffics of the periods brought into the comparison are to be duly considered. The following is a selection of a few principal streets in the City, with average numbers of vehicles by which they were traversed in 12 hours during one week-day, between 8 a.m. and 8 p.m., in July 1850, and in June, July, 1857. These data are abstracted from reports made by Colonel Haywood:—

TABLE No. 22.—CITY OF LONDON:—NUMBER OF VEHICLES WHICH TRAVERSED FIFTEEN PRINCIPAL THOROUGHFARES IN 12 HOURS, IN 1850 AND 1857; WITH A FEW DATA FOR 1865 AND 1871.

Name of Thoroughfare.	Number of Vehicles which traversed the street in 12 hours.		In 24 hours.	
	year 1850.	year 1857.	year 1865.	year 1871.
Temple Bar	7,741	9,883	11,972	—
Ludgate Hill	6,829	10,626	—	—
Cheapside	11,053	13,512	—	11,900
Poultry	10,274	11,667	—	9,600
Cornhill	4,916	5,256	—	—
Old Broad Street	—	—	—	2,600
Carried forward				

TABLE No. 22—*continued.*

Name of Thoroughfare.	Number of Vehicles which traversed the street in 12 hours.		In 24 hours.	
	year 1850.	year 1857.	year 1865.	year 1871.
Brought forward				
Leadenhall Street	5,930	4,325	—	—
Lombard Street	2,228	1,544	—	2,600
Fenchurch Street	3,642	5,273	—	—
Bishopsgate St. Within	4,842	6,283	—	—
Bishopsgate St. Without	4,110	5,804	7,366	—
Gracechurch Street	4,887	5,267	—	—
London Bridge	13,099	18,179	19,405	—
Newgate Street	6,375	8,341	—	7,400
Aldersgate Street	2,590	2,719	3,936	—
Blackfriars Bridge	5,262	6,723	9,660	—
Southwark Bridge	—	—	4,700	—
Totals	93,778	115,402		

NOTE TO TABLE.—Equestrians were included and numbered as vehicles. These numbers were so small as not materially to affect the quantities.

The gross increase of traffic, from 93,778 vehicles in one day in 1850, to 115,402 vehicles in one day in 1857, amounted to 22½ per cent. in seven years, or at the rate of about 3 per cent. per year, on the traffic observed in 1850. But, it must be remarked that the traffic may not be similarly augmented in every street individually, since the streams of traffic are modified by the opening of new streets and new railway stations. Instances of such fluctuations are observable in the numbers of vehicles given in the last column of the table.

The following data for the duration of 3-inch granite pavements in the streets of the City, have been gleaned from Colonel Haywood's reports. It is shown that the paving only lasted 6 years in the Poultry; and it may here be noted that the granite pavements of the Poultry, next to the kerbs, or next to the tramways, needed repair at about the end of the first year. Omitting the very short and very exceptionally tried pavement of the Poultry, as well as the second pavement in Cheapside, which was prematurely removed, the average of the observed durations was 15½ years:—

TABLE 23.—DURATION OF GRANITE PAVEMENTS IN SOME PRINCIPAL STREETS IN THE CITY OF LONDON. 1845—1863.

Aberdeen Granite Sets, 3 inches wide and 9 inches deep.

Situation.	Laid.	Relaid.	Taken up.	Duration.	Observations.
	Year.	Year.	Year.	Years.	
Poultry	1846	—	1852	6	Cost for repair, 2·66d. per square yard per year.
Cheapside (portion) {	1846	1853	1861	15	Taken up for asphalte.
	1861	—	1870	9	
Cheapside (east end)	1847	1853	—	—	
St. Paul's Churchyard	1847	1853 & '58	1863	16	
Ludgate Hill	1844	1853	1863	19	
Ludgate Street	1847	1853	—	—	
Fleet Street	1846	1853	1860	14	Cost for repair, 3d. per square yard per year.
Newgate Street	1846	1854	—	—	
Skinner Street	1845	1848 & '56	—	—	Widened in 1848.
Threadneedle Street (West)	1848	1857	1862	14·	
Fenchurch Street (East)	1846	1852	—	—	
Fenchurch Street (West)	1845	1852	1861	16	
Leadenhall Street {	1845	1852	1857	12	Relaid with one-third new stone.
	1857	1863	—	—	
Princes Street	1850	1857	—	—	Sets 4 in. × 9 in.

Average duration, omitting the Poultry and the second pavement in Cheapside as exceptional instances, 15¼ years.

To the above data for streets, may be added the following data for bridges :—

CONTINUATION OF TABLE No. 23.

Situation.	Laid.	Relaid.	Taken up.	Duration.	Observations.
	Year.		Year.	Years.	
London Bridge	1830	—	1842	12	6-inch sets.
	1842	—	1851	9	
Blackfriars Bridge	1840	—	1853	13	

In 1854, Colonel Haywood had estimated the total duration of the granite pavement of Fleet Street at 12 years. It may be seen from the table, that it lay actually 14 years till it was removed. He added that, after having been re-dressed, it would last 15 years more in secondary streets. The total life of such pavement amounted thus, by estimation, to 29 years.

The cost for repairs of Fleet Street was stated, in 1854, to be 3*d.* per square yard per year. The average cost of repairing the granite pavements of the streets of the City, in 1854, was also 3*d.* per square yard per year.*

Colonel Haywood, in 1853, made a careful estimate, based on the experience of City pavements, of the cost and the duration, or life, of a pavement of 3-inch Aberdeen granite sets, 9 inches deep, laid in such a thoroughfare as Gracechurch Street, in which the traffic as indicated in the table, No. 22, was rather below the average of the principal streets. This estimate was based on the prices of 1842, in which year he assumes the laying of the pavement for the purpose of a direct comparison with the cost of Carey's wood pavement in Gracechurch Street. The cost of the first 3-inch Aberdeen granite sets laid in the City was 14*s.* 6*d.* per square yard, laid and grouted complete, exclusive of foundation. The duration was taken at 25 years, as from 1842,—assuming that the granite was not

* Stated by Colonel Haywood.—See *Proceedings of the Institution of Civil Engineers*, vol. xiii. page 231.

GRANITE PAVEMENT, THREE-INCH SETS. 185

removed until it was quite worn out. During this time, it would have required three general relays, at a cost of 1s. per square yard, in addition to ordinary repairs, costing ¾d. per square yard per year, for 20 out of the 25 years of life, allowing the first five years free of cost for repairs:—

	Per sq. yd. for 25 years. s. d.	Per sq. yd. per year. d.
First cost, excluding foundation	14 6	6·96
Repairs; three relays at 1s.	3 0 }	2·04
Ditto, 20 years at ¾d. per year	1 3	
Total Expenditure	18 9	9
Deduct value as old material	2 3	1·08
Net total cost	16 6	7·92*

This estimate was formed on the assumption that the stone would have been worn out where it was first laid. In actual practice, it would have lain there from 14 to 18 years, and would have been removed to streets of inferior traffic, where it would have lasted at least as long again. The total life would thus have reached to from 30 to 40 years; and this may be taken as the actual life of 3-inch granite sets under the given conditions, with a value as old material of from 1s. 6d. to 2s. 6d. per square yard. If the whole life, as for Gracechurch Street, be taken as 35 years, it will appear, from a calculation similar to the above, that the net total cost is 7d. per square yard per year.

London Bridge was considered by Colonel Haywood to be the busiest thoroughfare in the world. In the course of 12 hours, in one day in 1850, it was traversed by 13,099 vehicles. The whole surface of the bridge was covered with a bed of clay, 15 inches thick, thoroughly puddled and well beaten together. Upon this a 3-inch layer of

---

* Colonel Haywood makes out the cost per square yard per year at about 7¾d.

fine sand was laid; next, a bed, 12 inches deep, of fine flint stones broken into small pieces, not larger than 2 inches in diameter, well dressed and rolled. Finally, it was paved with granite sets 6 inches by 9 inches deep, and was opened in 1830. After a period of 12 years, it was, in 1842, replaced by new paving, when the old stone was sold for 3s. per square yard. The second paving, laid in 1842, was formed of 3-inch Aberdeen granite sets, 9 inches deep; it was down 9 years, and was taken up and replaced with new granite in 1851. The second paving, though lifted, was not worn out; the average wear of the stones was estimated, from the result of careful observation, at first 2 inches of depth,—being, at the rate of ·222 inch per year; and, had they been dressed, sorted, and relaid, with a small quantity of new stones to supply the deficiency of those which were very much worn, or badly cut, they would have lasted 7 or 8 years more upon the bridge, and would then have been fit to lay in a secondary thoroughfare. They would have been relaid on the bridge, but for the fact that more time would have been consumed by this process, than by repaving with new material.

The total area of the carriage-way of London Bridge is 3,950 square yards. The following were the expenses actually incurred on the pavement, from 1842 to 1851:—

	For 9 years.			Per sq. yd. for 9 years.		Per sq. yd. per year.	
	£	s.	d.	s.	d.	s.	d.
First cost	3,850	16	0	19	6	2	2
Repairs for 9 years	277	11	0	1	4·86		1·87
Total expenditure	4,128	7	0	20	10·86	2	3·87
Deduct allowance for old stones	757	0	0	3	10		5·11
Net total cost for 9 years	3,371	7	0	17	0·86	1	10¾

The cost of the substratum is not included in this statement of expenditure. The high first cost was occasioned by, the great difficulty experienced in obtaining 3-inch stones, which had but recently been introduced, unless a high price was paid for them; together with the forcing of the labour, and the performance of the work by a half-width at a time.

The paving of Blackfriars Bridge, in 1840, with 3-inch sets has already been noticed, page 173. The paving consisted of Aberdeen stones and Guernsey stones, and was done at a cost of 20s. per square yard, including a foundation of concrete 12 inches deep. The average amount of wear, after 13 years' work, were 1½ inches of the Aberdeen granite, and ¼ inch of the Guernsey granite.*

In 1871, Colonel Haywood made estimates, based upon past experience, of the durability and cost, under not unfavourable circumstances, of Aberdeen granite pavement, 3 inches wide and 9 inches deep, in Cheapside, Poultry, Old Broad Street, Moorgate Street, and Lombard Street, which, varying in width as well as in the nature of the traffic, might be taken as types of the leading thoroughfares of the City. The cost of pavements generally had augmented gradually during the previous 30 years. In 1854, the cost was from 15s. to 17s. per square yard. In 1871, the price of paviors' work, above described, complete, except the foundation, was 15s. 4¼d. per square yard; but he adopted, in his estimates, a cost of 16s. per square yard.

The estimated duration of a granite pavement in Cheapside was 15 years; and assuming one entire relay, and allowing for the value of the old stone when removed at the end of the time, the total cost during the period of 15 years would be £1 4s. 4½d., or 1s. 7½d. per square yard, per year.

Similarly, for the Poultry, Old Broad Street, Moorgate

* *Proceedings of the Institution of Civil Engineers*, vol. xiii. page 237.

Street, and Lombard Street, suitable periods of duration were assigned, and one general relay during the time, with allowance for the value of the old stones. The following table comprises particulars of Colonel Haywood's estimates for the several streets:—

TABLE NO. 24.—ESTIMATED DURATION AND COST OF GRANITE PAVEMENTS IN PRINCIPAL CARRIAGE-WAYS IN THE CITY OF LONDON, 1871.

Aberdeen Granite Sets, 3 inches wide by 9 inches deep.

Situation.	Width of Street.	Daily traffic in 24 hours	Duration of the Pavement.	First cost per square yard.	Total cost, including maintenance, and deducting for old stones.	Average total cost per square yard per year.
	Feet.	Vehicles.	Years.	s. d.	s. d.	s. d.
Cheapside	30	11,900	15	16 0	24 4½	1 7½
Poultry	22	9,600	8	16 0	22 4	2 9½
Old Broad Street	24	2,600	20	16 0	20 11¼	1 0¼
Moorgate Street	32	7,400	15	16 0	20 7	1 4¾
Lombard Street	17	2,600	20	16 0	21 4½	1 0¾
Averages	—	—	15·6	16 0	21 11	1 7

The average estimated duration indicated by this table is nearly identical with that which is deduced from actual observation in the previous table, No. 23.

Typical sections and plans of a 50-feet street for the City of London, Figs. 46, 47, and 48.—For these illustrations the Author is indebted to Colonel Haywood. The extreme width of the street is 50 feet, between the houses:—divided into 30 feet for the width of the carriage-way, and 10 feet for each footway. The bed of the road is excavated to a depth of 21 inches below the finished level of the street, following the contour of the surface. A layer of broken stones, 9 inches thick, is distributed over the ground, and is covered by a layer of hoggin or small gravel and sand, 3 inches thick, as a bed for the paving.

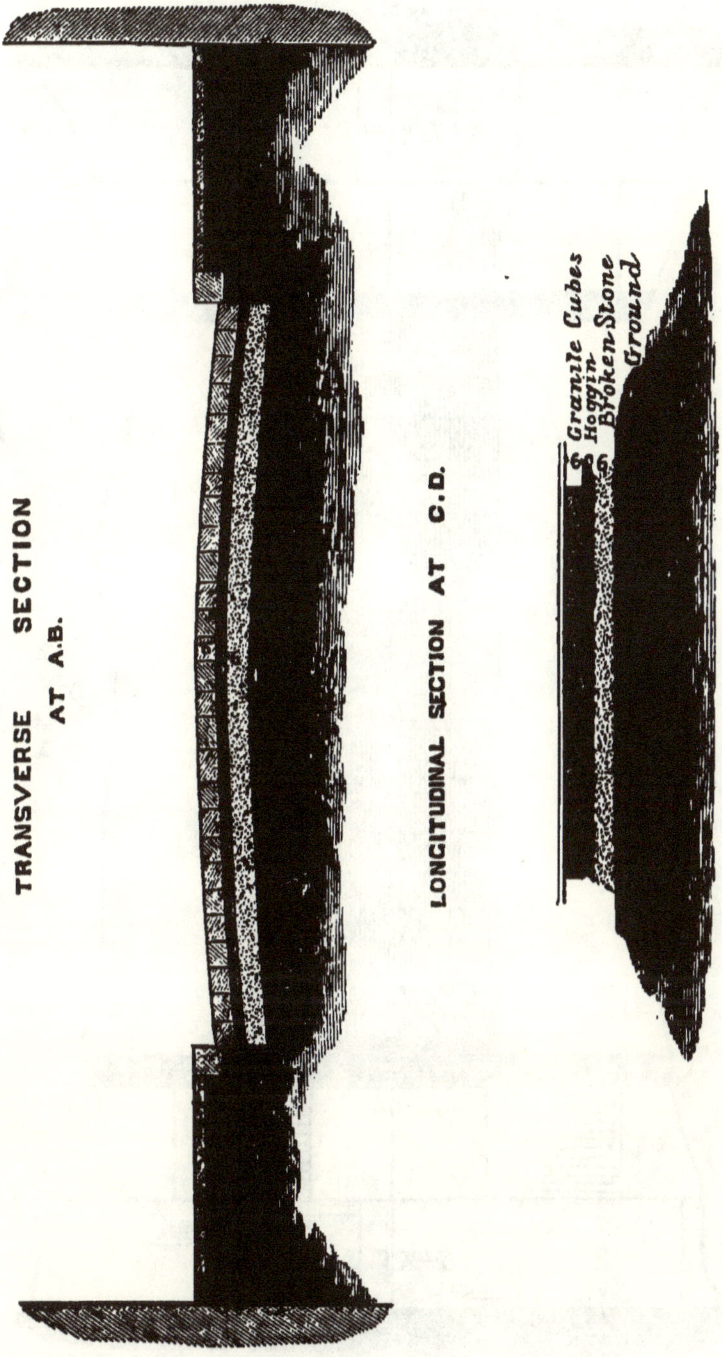

FIFTY-FEET STREET. 189

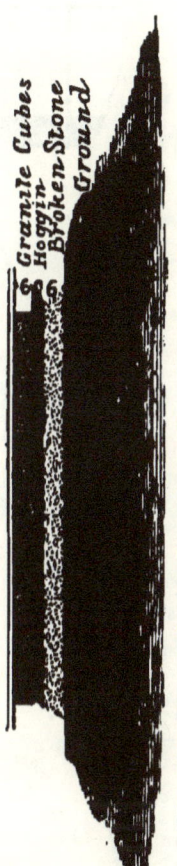

TRANSVERSE SECTION AT A.B.

LONGITUDINAL SECTION AT C.D.

Granite Cubes
Hoggin
Broken Stone
Ground

Figs. 46 and 47.—Fifty-feet Street in the City of London.—Sections.

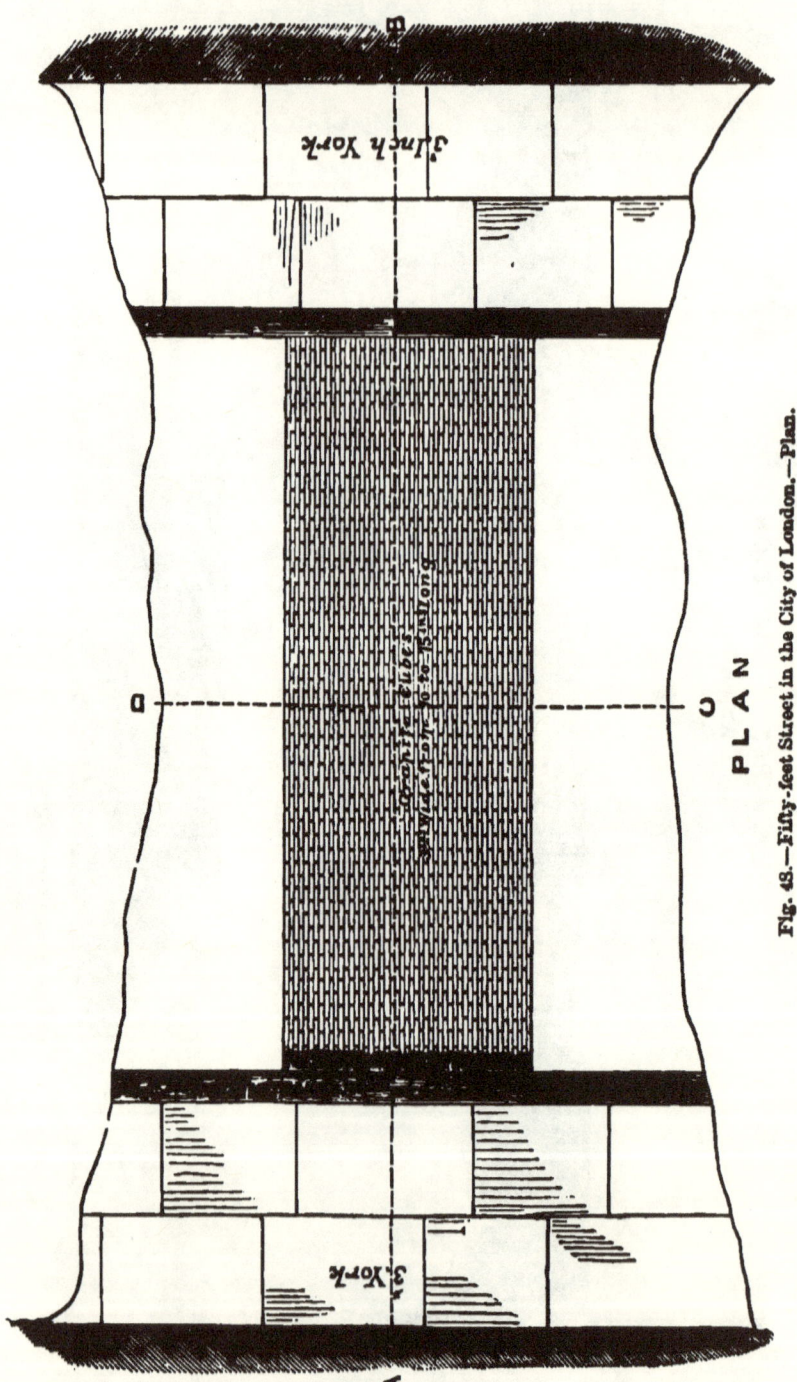

Fig. 43.—Fifty-feet Street in the City of London.—Plan.

The paving consists of granite sets or "cubes," 3 inches wide and 9 inches deep, and of lengths varying from 10 to 15 inches, grouted at the joints. The rise of the pavement is 6 inches for the width of 30 feet, or 1 in 30 for the average inclination, the contour being a segment of a circle. The footpaths are laid with 3-inch York pavement, bounded by a granite kerb 12 inches wide, and 9 inches deep, showing 6 inches above the roadway pavement.

Southwark Street, Southwark, Figs. 49 and 50.—This is a good example of a first-class Metropolitan street, arranged with a subway and a sewer at the middle, and cellarage at each side. The street is 70 feet wide between the houses, comprising two 12-feet footways, and a carriage-way 46 feet wide. For the construction of this street, the ground was levelled, and the soft places cleared out. It was covered with a bottoming of brick rubbish, varying from 6 to 10 inches deep, which was rolled and bound with sand. Upon this bottoming was laid a stratum of concrete

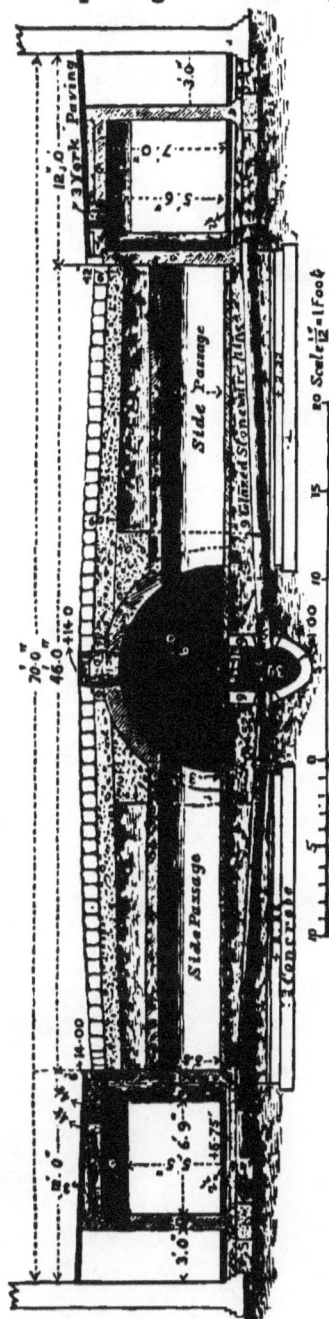

Fig. 49.—Southwark Street, Southwark. Transverse Section.

12 inches thick, consisting of blue lias lime and clean Thames ballast, in the ratio of 1 to 6 by measure. A

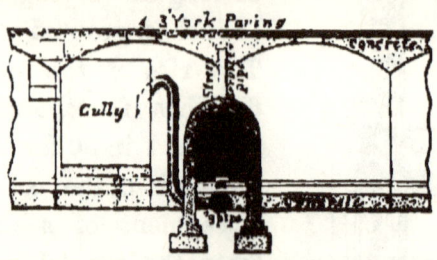

Fig. 50.—Southwark Street, Southwark.—Longitudinal section under footpath.

layer of sand or hoggin, 1½ or 2 inches thick, was distributed over the concrete as a bed for the granite sets, which were 9 inches deep and 3 inches wide. The stones were set close and grouted together.

CHAPTER VIII.

STONE PAVEMENTS OF LIVERPOOL.

According to the report of Mr. Newlands, the borough engineer of Liverpool, in 1851, there were 174 miles of carriage-way, and 69 miles of courts and passages, comprising—

	Square yards.	
Carriage-ways	2,243,560	
Channels	231,362	
		2,474,922
Footways		1,048,264
		3,523,186 square yards.

From this it appears that the average width of the carriage-ways and channels together was 8·1 yards, or 24·3 feet. The covering of the carriage-ways consisted for the most part,—to the extent of nearly two-thirds,—of boulder paving; one-fifth consisted of macadam, 8 per cent. only consisted of pavement of greywacké sets from Penmaenmaur, and 6 per cent. consisted of sand and ashes.

The actual cost for the construction of several streets with greywacké pavement, in 3-inch sets, is given in Table No. 25. The channels were of Penmaenmaur stone, and the crossings of granite. The cost for the preparation of the foundation is included. The foundation consisted of a bottoming of ballast prepared and covered with gravel and sand, upon which the sets were placed.

K

A similar table, No. 26, gives the cost for the construction of several streets with boulder paving; from which it may be seen that the boulders were bedded on the ground, without the intervention of a hard bottoming.

TABLE No. 25.—LIVERPOOL STREETS:—COST FOR CONSTRUCTION PER SQUARE YARD OF THE CARRIAGE-WAYS OF SIX STREETS IN DIFFERENT LOCALITIES.

Greywacké Set Pavements.—Sets, 3 inches wide, 7 inches deep.

	Name of Street.						Average First Costs.
	Strand.	Temple.	Crosby.	Grafton.	Rumford.	Elliot.	
MATERIALS.	pence.	pence.	pence.	pence.	pence.	pence.	pence.
Stone Sets	37·14	29·47	24·72	36·26	38·72	26·92	32·20
Gravel, Sand, Ballast	5·38	5·90	8·23	·89	2·03	2·08	4·09
Lime	1·23	2·28	1·19	1·08	2·65	1·01	1·57
Crossings	3·14	2·75	2·28	·03	·67	10·07	3·16
Channels	1·60	6·69	2·58	3·18	4·40	—	3·07
Materials	48·49	47·09	39·00	41·44	48·47	40·08	44·09
Cartage	13·60	8·26	2·34	4·04	1·42	1·97	5·27
Total Cost for Materials	62·09	55·35	41·34	45·48	49·89	42·05	49·36
LABOUR.							
Paviors	5·59	6·92	4·69	1·93	3·29	} 7·89	{ 4·19
Labourers	11·43	12·23	9·58	4·84	10·38		{ 8·93
Labour	17·02	19·15	14·27	6·77	13·67	7·89	13·12
Use of Tools	·90	1·33	·51	·45	·72	·39	·72
Total cost for Labour	17·92	20·48	14·78	7·22	14·39	8·28	13·84
Total Cost for Materials & Labour	80·01	75·83	56·12	52·70	64·28	50·33	63·20
or	6s. 8d.	6s. 3¾d.	4s. 8d.	4s. 4¾d.	5s. 4¼d.	4s. 2¼d.	5s. 3¼d.

TABLE No. 26.—LIVERPOOL STREETS:—COST FOR CONSTRUCTION PER SQUARE YARD OF THE CARRIAGE-WAYS OF SIX STREETS IN DIFFERENT LOCALITIES.

Boulder Pavement.

	Name of Street.						Average First Costs.
	Spencer	Hygeia.	Sheridan.	Jubilee.	Hope.	Hope Place.	
MATERIALS.	pence.	pence.	pence.	pence.	pence.	pence.	pence.
Boulders	14·61	17·52	12.28	13·06	9·76	11·25	13·08
Channels	4·08	3·96	9·82	6·39	4·51	6·80	5·93
Crossings	·27	1·88	13·60	8·51	3·35	1·66	4·88
Gravel and Sand	—	—	·22	·61	·41	·69	·32
Materials	18·96	23·36	35·92	28.57	18·03	20·40	24·21
Cartage	2·83	3·76	8·34	5·49	3·06	1·21	4·11
Total Cost for Materials	21·79	37·12	44·26	34·06	21·09	21·61	28·32
LABOUR.							
Paviors	3·51	3·24	2·83	4·17	2·88	3·80	3·40
Labourers	5·27	4·86	11·74	4·98	3·30	4·53	5·78
Labour	8·78	8·10	14·57	9·15	6·18	8·33	9·18
Use of Tools	·51	·48	·98	·37	·62	·46	·57
Total Cost for Labour	9·29	5·58	15·55	9·52	6·80	8·79	9·75
Total Cost for Materials & Labour,	31·08	35·70	59·81	43·58	27·87	30·40	38·07
or		2/11¾	4/11¾	3s.7½d.	2s. 4d.	2s.6½d.	3s. 2d.

These tables afford good selections of instances to show how much the cost is influenced by the circumstances under which the work is done. The cost of the materials used in set-paving does not vary so much as the cost of cartage and of labour, which depend upon the distance of a street from the docks whence the material is conveyed; and upon the length of the day and the state of the weather, and other circumstances affecting the convenience for operations.

From these tables it appears that the average costs for the construction of the carriage-ways of Penmaenmaur sets and of boulders, were composed as follows:—

	Liverpool.	Penmaenmaur sets.		Boulders.	
		s. d.		s. d.	
Materials	.	4 1¼	or 78 per cent.	2 4¼	or 74 per cent.
Labour	. .	1 2	or 22 ,,	0 9¾	or 26 ,,
Total cost	.	5 3¼	100	3 2	100

In introducing a comparative statement of the annual cost for maintenance of set-paving, boulder-paving, and macadam, Mr. Newlands fixed the first cost—

	Liverpool.	First cost.
For square-set paving,	at 4s. 6d.	per square yard.
,, boulder paving	,, 1s. 8d.	,,
,, macadam	,, 1s. 8d.	,,

These amounts were probably based on wider averages than those which have been deduced from the foregoing tables. However that may have been, Mr. Newlands, in 1854, stated that, at Liverpool, the first cost for a set-paved road was 6s. per square yard, and that of a macadam road was 2s. per square yard.*

The average cost for maintenance for the three years 1848—50, is given for 27 streets, of which nine were paved with squared sets, nine with boulders, and nine were laid with macadam. Summaries of the costs are given in the table, No. 27.

Mr. Newlands explained that the averages here derived were not to be taken as for long periods,—especially for the boulder paving, much of which was required shortly to be

* *Proceedings of the Institution of Civil Engineers*, vol. xiii., 1853–54, page 240.

STONE PAVEMENTS OF LIVERPOOL.

TABLE NO. 27.—LIVERPOOL STREETS:—ANNUAL COST FOR MAINTENANCE OF 27 STREETS, DEDUCED FROM THE AVERAGED CHARGES FOR THREE YEARS, 1848–50.

Penmaenmaur set paving.		Boulder paving.		Macadam.	
Street.	Annual cost per square yard.	Street.	Annual cost per square yard.	Street.	Annual cost per square yard.
	pence.		pence.		pence.
St. James'	·08	Blundell	·55	Church	8·60
Park Lane	·29	Mersey	·79	Berry	8·33
Whitechapel	·11	Flint	·73	Renshaw	18·31
Dale	·38	Simpson	1·95	Gt. George	3·81
Castle	·17	Sparling	·37	Park Lane	16·29
North John	·26	Limekiln Lane	·62	Myrtle	5·82
Marybono	·09	Burlington	1·14	Hanover	14·61
Water	·19	Soho	·18	Gt. Howard	12·54
Leeds	·06	Fox	1·01	St. Anne	3·80
Averages	·18		·82		10·23

renewed. He deduced from the tables that, when the macadam was subjected to heavy loads, the annual cost of maintaining it was nearly equal to the cost of renewing it, and that the vertical wear amounted to 12 inches per year. Where the loads were light, although the traffic was great, the wear of the macadam was only about half as great as under heavy loads; and that, under the same circumstances of traffic, the expense was reduced when the street was wide and open, and the moisture was rapidly evaporated by the action of the sun and the air. He concluded from the experience of Castle Street, Dale Street, and Marybone, that set-paving would last 20 years before requiring to be renewed. He decided, finally, that the total annual costs of maintenance were respectively,

LIVERPOOL.   Maintenance annually.
For square-set paving . . ·28 pence per square yard.
 „ Boulder paving . . . 6·80   „   „
 „ Macadam . . . . 16·00   „   „

CHAPTER IX.

STONE PAVEMENTS OF MANCHESTER.

The early pavements for the streets of Manchester consisted of boulder stones brought from the sea-coasts of Wales, Westmoreland, and Cumberland, like those of Liverpool, which have already been described. But the construction of boulder pavement has long since been discontinued; the importation of boulders into Manchester ceased about the year 1840. There yet remains a considerable area of such pavement, particularly in the older streets of small traffic, in the outskirts. They are in process of renewal with granite sets. When finely broken, the boulder stones make an excellent "racking" or packing for paving sets.

The principal thoroughfares, according to Mr. H. Royle, District Surveyor,* are paved chiefly with syenitic granite, or with trap-rock sets. The most usual dimensions of the sets are 5 inches, 6 inches, and 7 inches in depth; from 3 to $3\frac{3}{4}$ inches in width, and from 5 to 7 inches long. The largest and deepest sets are laid in streets having the heaviest traffic. Cubes of 4 inches were tried, some years ago; but it appeared that they were unfitted for resisting the lateral stress of the traffic,—particularly on streets of considerable inclination. Stones are obtained from, 1st. The Welsh Granite Company, and the Portnant Granite

---

* See a paper "On Street Pavements in Manchester," by Mr. H. Royle, District Surveyor; read July 9, 1876, before the Association of Municipal and Sanitary Engineers and Surveyors.

STONE PAVEMENTS OF MANCHESTER.

Company, near Carnarvon. 2nd. The Syenite Granite Company, Portmadoc. 3rd. From greywacké quarries at Penmaenmaur, North Wales. 4th. From quarries at Clee Hills, Shropshire. 5th. From Newry, Ireland. It is said that the prices of sets from these quarries are nearly the same.

The secondary streets, of little traffic, are paved with sets of millstone-grit, brought from the mountainous districts of Lancashire and Derbyshire.

Macadam, which is regarded in Manchester as an expensive nuisance, is being gradually replaced with granite pavement.

The paved carriage-ways are constructed in the following manner:—A foundation, not exceeding 15 inches in thickness, is laid, of cinders and other hard material, including 3 inches of gravel as a bedding for the sets. The traffic is turned over this foundation until it becomes solid; and the temporary gravel surface is renewed from time to time. When the surface has become sufficiently solid, the sets are bedded upon it, and well beaten, and they are racked with clean small broken stones, or with washed gravel. The joints of the stones are then filled up with an asphaltic fluid mixture composed of coal-tar, pitch made from coal-tar, gas-tar, and creosote oil; in the proportion of 1 cwt. of pitch to 4 gallons of tar, and 1 gallon of creosote,—proportions which are varied somewhat according to the quality of the pitch employed. The mixture is melted and boiled for from 1 to 2 hours, in a boiler adapted for the purpose, before it is poured into the joints.

The use of the pitch-composition for closing the joints of the stone-sets was originally suggested by Mr. Ronchetti, a chemist, of Manchester; and the good reputation of Manchester pavement is, in a great measure, due to the employment of this compound for jointing, together with the solid foundation prepared for the pavement. The jointing is impervious to moisture, and possesses a degree of elasticity sufficient to prevent it from cracking. It adapts itself to all temperatures. Sometimes, as Mr. Royle observes, in the heat of the summer, the asphalte, as he calls it, rises out of the joints, and slowly flows to the channels; but this, it is said, rarely happens, and it happens only when the asphalte has not been properly prepared. It follows that the foundation is always dry, and as no material rises from below, the formation of mud is prevented, and the cleansing of the surface is easily done. The foundations of hard core, prepared in the manner described, have given entire satisfaction; and there has not been any need for the employment of concrete in foundations.

The pitch obtained for the jointing of the pavement has, during the last 10 or 12 years, deteriorated in value for the purpose,—since the demand for gas-tar for the manufacture of aniline and other dyes, which has greatly increased,—by the removal of that ingredient in virtue of which the asphalte retained its elasticity and adhesiveness. The deficiency is partially supplied by the addition of the gas-tar, which is added to the pitch whilst it boils. Still, it is said, the asphalte laid in the old pavements was much better than that which is now made; and the carriage-way pavements more recently constructed are not likely to last so long in good condition as the older asphalted pavements.

The cost for repaving an old street with 6-inch sets, exclusive of the cost for foundation, and without taking credit for old material, is as follows:—

For stone sets, 6 inches wide . .	8s. 4d.	per square yard.
Labour, carting, gravel, asphalte .	2s. 6d. to 2s. 9d.	,,
Total	10s. 10d. to 11s. 1d.	,,

The amount of the second item for labour, &c., is a minimum in summer, when the men work for 54½ hours per week; and is a maximum in winter, when the hours of labour are fewer. The average cost for the year is 11s. per square yard.

Mr. Royle states that the cost for maintenance of asphalte-granite pavement is in simple proportion to the density of the traffic; at the centre of the City, the cost is greatest. He adds that the average duration of the pavement, as first laid, is 14 years; at the end of which time it requires to be entirely relaid; and that a sinking fund of 3d. per square yard per year would provide labour and material for maintenance in perpetuity.

Macadam in Manchester.—The cost of macadamising Albert Place, Manchester, in 1852, exclusive of foundation, comprising an area of 1,373 square yards, was as follows:—

		Per sq. yd.
		s. d.
606 tons 9 cwt. of macadam at 10s.	£303 4 6	4 5
Labour	11 18 8	0 2
Carting	24 15 4½	0 4¼
Total	339 18 6½	4 11¼

Taking the weight of one cubic yard of macadam at 1⅛ tons, the bulk of the macadam used was (606·5 × $\frac{8}{9}$ =) 539 cubic yards, which was equivalent to a layer 9 inches in thickness. The cost of maintenance, including scavenging, was 1s. 1½d. per square yard per year.

CHAPTER X.

WEAR OF GRANITE PAVEMENTS.

It has been stated that the greywacké sets from Penmaenmaur are practically limitless in their resistance to wear; and, as a corollary, they make dangerously slippery pavements. Next to these, the sets of Guernsey granite may be ranked. Mount Sorrel granite comes after; and Aberdeen granite sets, the most generally popular, are the softest and the least durable; that is, they wear the most rapidly. The order of resistance to wear of other stones, —the results of Mr. Walker's experiments,—has been given, page 126.

The data for the vertical wear of granite pavement, forming the basis of the following table, No. 28, are collected from various sources*:—

TABLE No. 28.—WEAR OF CARRIAGE-WAY PAVEMENTS IN THE CITY.
Aberdeen-granite stones.

Locality.	Year of observation.	Width of stones.	Number of years down.	Amount of vertical wear.		Number of years to wear away 1 inch.
				Total	per Year.	
		Inches.	Years.	Inches.	Inch.	Years.
Watling Street	1844	6	17	$1\frac{1}{32}$	·065, or $\frac{1}{16}$	15·5
Great Tower Street	?	6	9	$1\frac{1}{4}$	·139, or $\frac{1}{7}$ nearly	7·2
Bishopsgate Street Without	1850	6	20	$2\frac{3}{32}$	·105, or $\frac{1}{10}$ fully	9·6
St. Paul's Churchyard	1847	6	16	2	·125, or $\frac{1}{8}$	8
Fleet Street	1846	6	14	2	·143, or $\frac{1}{7}$	7
Poultry	1852	3	6	$1\frac{1}{4}$ to $1\frac{1}{2}$	·230, or $\frac{2}{8}$	4·4
London Bridge	1851	3	9	2	·222, or $\frac{2}{8}$	4·5
Blackfriars Bdg.(old)	1853	3	13	$1\frac{1}{2}$	·115, or $\frac{1}{8}$	8·7
Do.(Guernsey stones)	1853	3	13	$\frac{1}{4}$	·019, or $\frac{1}{52}$	52

* Chiefly from the *Proceedings of the Institution of Civil Engineers*, vol. ix., page 214; and vol. xiii., page 221.

It may be expected that there is a correspondence between the intensity of the traffic and the rate of vertical wear of the several granite pavements. But, in forming such a comparison, there is another factor to be introduced, —the width of the pavement; for the vertical wear may be expected to be directly as the quantity of daily traffic, and inversely as the width of the street; and the quotient of the traffic by the width, or the number of vehicles per foot or per yard of width, measures the intensity of the traffic. The following table, No. 29, has been constructed to show to what extent the relation holds good for such of the thoroughfares as have had their traffic observed. The traffic, showing the number of vehicles that traversed each locality in ten hours in one day, in 1850, already given in a previous table, is calculated for one foot of width of each locality; and the last column contains the products of the relative intensities of traffic, in column 4, by the periods of year required to wear the pavement through equal vertical depths of one inch. These products represent proportionally the relative amounts of traffic which caused equal amounts of vertical wear:—

TABLE No. 29.—WEAR OF CARRIAGE-WAY RELATIVE TO INTENSITY OF TRAFFIC.

Aberdeen Granite Stones.

LOCALITY.	Approximate average width of pavement.	Vehicles in one day in 1850.		Number of years to wear away one inch.	Relative amounts of traffic for equal amounts of vertical wear.
		Total.	Per foot of width.		
Sets 6 inches wide.	Feet.	Vehicles.	Vehicles.	Years.	Col. 4 × col. 5
Gt. Tower Street	16	2,890	181	7·2	1,303*
Bishopsgate St. Without	22 (narrowest part.)	4,110	190	9·6	1,824
St. Paul's Churchyard	33	6,829	207	8	1,656
Fleet Street	30	7,741	258	7	1,806

TABLE No. 29—*continued.*

| LOCALITY. | Approximate average width of pavement. | Vehicles in one day in 1850. | | Number of years to wear away one inch. | Relative amounts of traffic for equal amounts of vertical wear. |
		Total.	Per foot of width.		
Sets 3 inches wide.	Feet.	Vehicles.	Vehicles.	Years.	Col. 4 × col. 5
Poultry	22	10,274	467	4·4	2,055
London Bridge .	35	13,099	374	4·5	1,683
Blackfriars Bdg. (old)	28	5,262	188	8·7	1,636
1	2	3	4	5	6

Supposing that the intensities of traffic be truly measured by the quantities in the 4th column of the table, and that the wear had been exactly in the inverse ratio of the intensity of traffic, the relative products in the last column should have been equal to each other. But the quantities in the 4th column can be but roughly approximate to each other, for they are based on but one day's observations; besides, the traffic, though it was distinguished according to the number of horses to a vehicle, was not classified so as to distinguish goods traffic from passenger traffic. There is, nevertheless, a remarkable degree of correspondence in the products or measures of performance. The Poultry exhibits the highest product; and that is consistent with the fact that the traffic consisted of a less proportion of waggon traffic than that of London Bridge or Fleet Street, and that the traffic of the Poultry was limited to a walking pace. Again, though there is a much larger proportion of heavy traffic over London Bridge, yet the heaviest traffic moves at a walking pace, and thus the pavement holds a high position in the order of relative duration. It is remarkable, too, that Blackfriars Bridge, with an intensity of traffic exactly one half of that of London Bridge, shows, in column 5, about twice the actual duration; and that, consequently, the products in the last column are substantially identical. The performance of Bishopsgate pavement is high; and the highness of the rate may be partially attributable to the fact of the irregularity of width of the street, for here and there it expands into large bays which dilute the traffic to some extent at certain localities.

The evidence in the last table supports generally Colonel Haywood's conclusion, drawn from the results of his observations, that the duration of a pavement is "almost exactly" in the inverse ratio of the amount of the traffic over it.

From the foregoing materials, some useful general data may be adduced for the wear and duration of granite pavements. For the seven pavements named in Table No. 29, the traffic and the wear were as follows:—

Aberdeen Granite Pavements.	Vehicles per foot of width.	Vertical wear per year.
Great Tower Street	181	·139 inch.
Bishopsgate Street Without	190	·105 ,,
St. Paul's Churchyard	207	·125 ,,
Fleet Street	258	·143 ,,
Poultry	467	·230 ,,
London Bridge	374	·222 ,,
Blackfriars Bridge (old)	188	·115 ,,
Totals	1,865 vehicles.	1·079 inches.
Averages	266 ,,	·154 ,,

From these averages, it appears that the wear per 100 vehicles per day per foot wide, was equal to ($·154 \times \frac{100}{266} =$) ·058 inch, or about $\frac{1}{18}$ inch per year.

To apply this rate of wear to the representative streets in Table No. 24, page 188, the daily traffics in 12 hours are here taken, for the sake of uniformity of comparison:*—

* These traffics are taken from a table of Asphalte Pavements, in a following page.

Cheapside	298	vehicles per foot of width.
Poultry	363	,, ,,
Old Broad Street	103	,, ,,
Moorgate Street	188	,, ,,
Lombard Street	147	,, ,,
Average	220	

The average duration of the granite pavements in these streets is given in Table No. 24 as 15·6 years. Take it at 15 years, as a round number. Then, the total vertical wear of the pavements when removed is ($\frac{1}{16} \times 15 \times \frac{100}{100} =$) 2·06 inches, say 2 inches.

Again, the stones, after having been redressed, last 20 years longer, in minor streets, for which the traffic averages 150 vehicles in 12 hours, per foot of width. Then the total vertical wear in the minor streets is ($\frac{1}{16} \times 20 \times \frac{100}{100} =$) 1·875 inches, or, as a round number, 2 inches.

TABLE No. 30.—CITY OF LONDON:—RECAPITULATION OF DATA ON THE WEAR AND DURATION OF ABERDEEN GRANITE PAVEMENTS.

Sets, 3 inches wide; 9 inches deep.

Aberdeen Granite Pavements.	Vertical wear.	Duration.
	Inches.	Years.
Vertical wear per 100 vehicles in 12 hours per foot of width, per year	$\frac{1}{16}$	1
Total vertical wear in principal streets	2	15
Ditto additional ditto in minor ditto	2	20
Total vertical wear when laid aside	4	35
Remaining depth when laid aside	5	
Depth of new sets	9	

Here it is seen that 3-inch sets may be reduced by wear to 5 inches of depth, when they are treated as worn out, and are laid aside.

The depth of a stone-set should be proportioned to the width at the surface. For insuring steadiness and resistance to the vertical and oblique forces applied to it by the traffic, the depth of a set cannot be too great; but it may be too small, for the resistance to displacement becomes less as the depth is reduced in proportion to the width. A flagstone laid in a foot pavement is an extreme case in point: it can, if at all loose, be more easily tilted on its bed, by forces acting on it, which may act with leverage, than if it were one of a row of flags on edge placed side by side. For reasons of this nature, old paving stones, from 6 to 8 inches wide, are to be met with in London, having a depth of 11 or 12 inches. Modern granite sets, 3 inches wide, are made 7 or 9 inches deep; and if 2-inch sets were used, a depth of 6 inches would be sufficient for stability. Thence there is a saving in first cost by laying narrow sets.

CHAPTER XI.

STONE TRAMWAYS IN STREETS.

Mr. Walker laid down two granite tramways on the Commercial Road, in the east of London; they consisted of two uniform lines of "tram-stones," each 16 inches wide, and 12 inches deep, and 5 or 6 feet in length, placed at a suitable distance apart, to carry the wheels of vehicles. The interspaces are paved. Such tracks are laid in several narrow streets in the City of London, at Holyhead, and in other places. When they were first laid down in England, each wheel-track had kerbs at the outer sides; but in recent constructions these have been dispensed with, and vehicles may move freely from one part of the road to another. From the experiments of Sir John Macneil on the resistance of granite trackways, the resistance amounted to only $12\frac{1}{4}$ lbs., or 13 lbs., per ton of gross weight:—less than half the resistance of granite paving. On an incline of 1 in 20, it is reported, the resistance was 132lbs. per ton, against a resistance of 295 lbs. per ton before the tracks were laid. In Glasgow, a tramway consisting of cast-iron plates, 2 inches thick, 8 inches wide, and cast in lengths of 3 feet, is laid in Buchanan Street, on a gradient of 1 in 20, where it has been down for 40 years.

Granite tramways* are in general use in Northern

---

* For the particulars of Italian tramways, the writer is indebted to the valuable "Report on Stone Tramways in Italy," by Mr. P. Le Neve Foster, Jun., C.E., appended to the Report of the Committee of the Society of Arts on "Traction on Roads," June 25th, 1875.

STONE TRAMWAYS. 209

Italy, not only in the streets of the principal cities—as Turin, Milan, and Verona—but also in smaller towns, as Chivasso, Mortara, and Vigevano. They consist of two parallel lines of granite blocks, usually 24 inches wide, 8 inches deep, and 5 feet in length, bedded in a layer of sand. See Fig. 51. The lines are 28 inches apart, and the interspace, or footway for horses, as well as the other portions of the roadway, are paved with cobbles obtained from the Po, or from other rivers. These stones should be egg-shaped, with a maximum diameter of from 3¼ to 4¼ inches, and a depth of from 4¼ to 5½ inches. The roadway is usually formed with a slight inclination downwards towards the centre. By this arrangement, the space between the trams serves as a channel to receive the surface water, and is provided with stone gratings, placed at suitable intervals, as shown in Fig. 52, by which the water escapes into the sewers. The surfaces of the trams are slightly inclined towards each other, the inner edges being ⅜ inch lower than the outer edges; whilst the interspace is concave, having a versed sine or

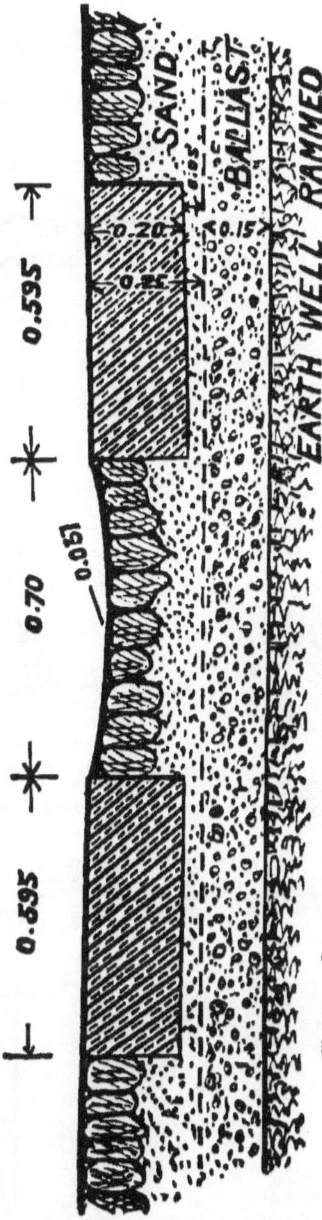

Fig. 51.—Granite Tramways in Northern Italy. Dimensions in Metric measure. Scale 1-20th.

210   STONE TRAMWAYS.

depression of 1½ inches. The foundation of the roadway

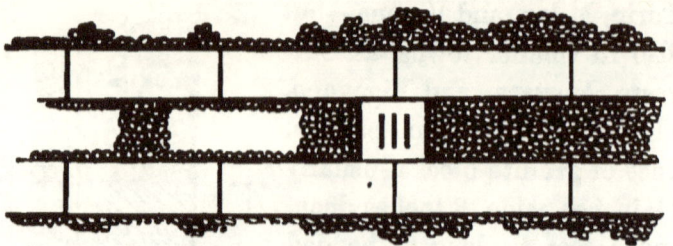

Fig. 52.—Plan of Tramway, Northern Italy, showing stone gulley.
Scale 1-100th.

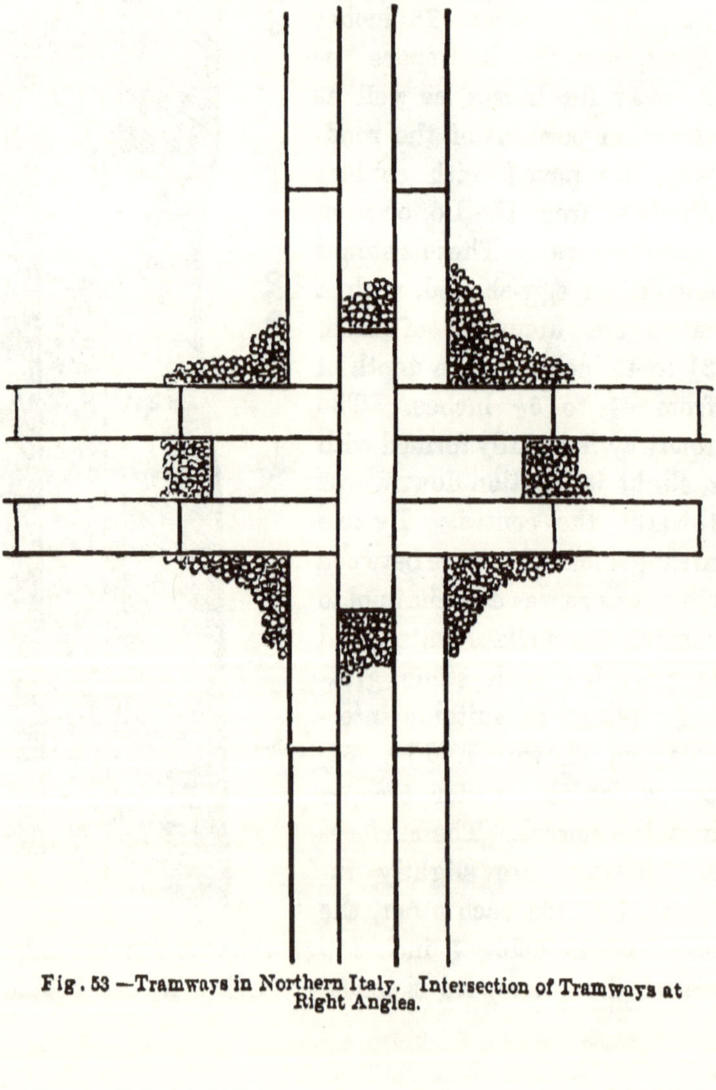

Fig. 53.—Tramways in Northern Italy. Intersection of Tramways at Right Angles.

consists of a layer of screened gravel, about 6 inches deep,

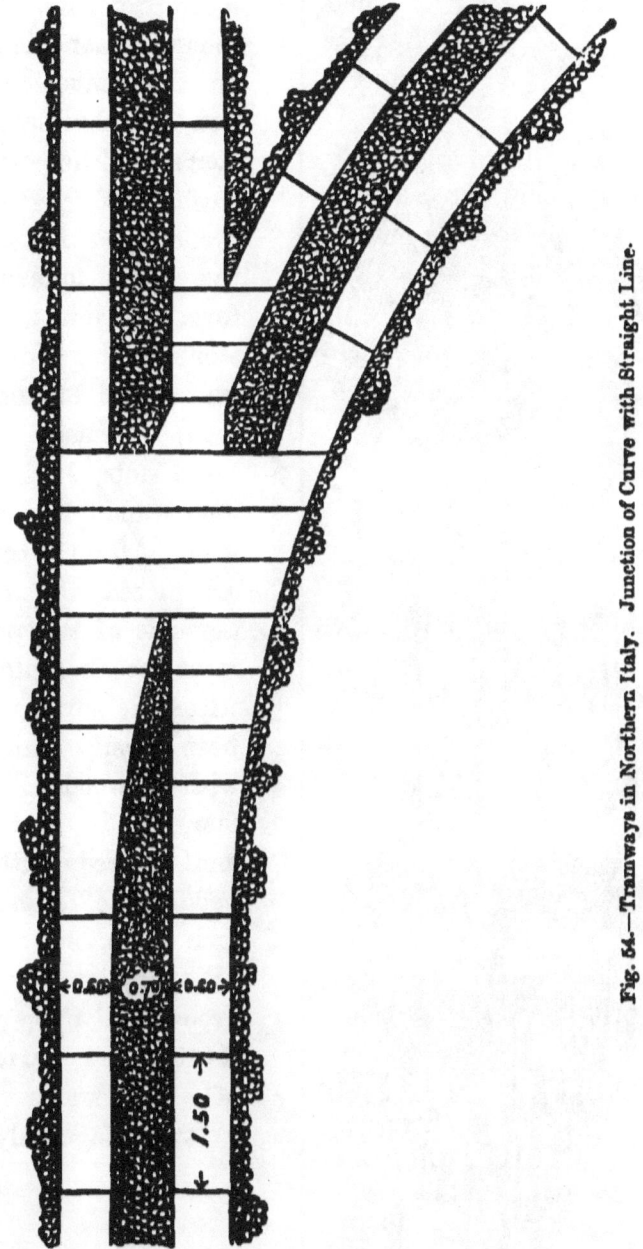

Fig. 54.—Tramways in Northern Italy. Junction of Curve with Straight Line.

placed on the surface of the ground, and well rammed and

212 STONE TRAMWAYS.

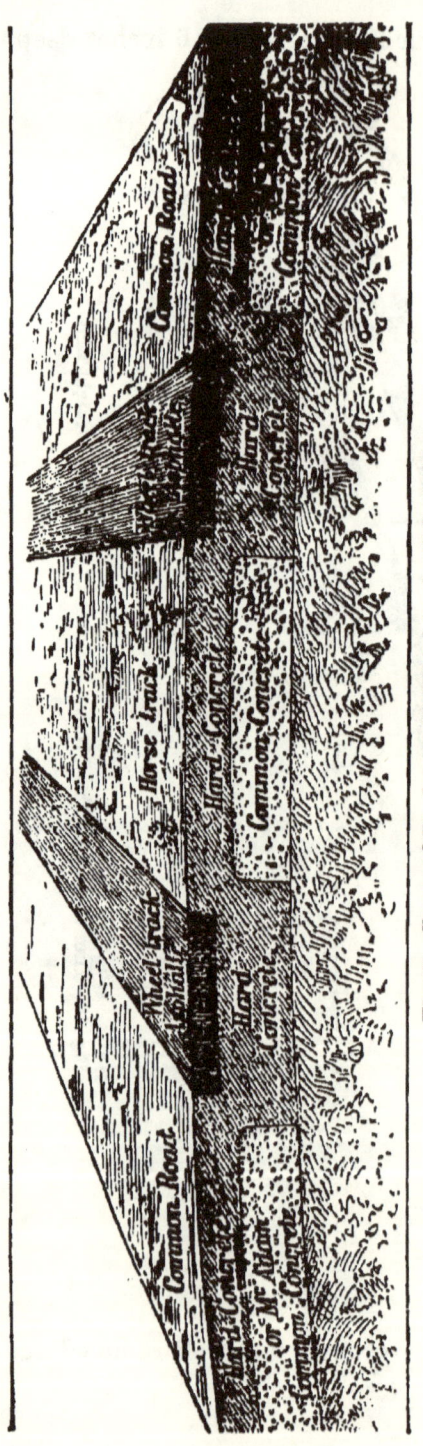

Fig. 65.—Proposed Roadway with Wheel-tracks of Asphalte.

watered, so as to form a compact mass. Two inches of sand is laid on the gravel, as a bed for the paving-stones. The upper surfaces of the trams are dressed flat, and the ends square, to form close joints. The stone gratings for the gullies are 32 inches long, formed with three slots, 12 inches long and 1¼ inch wide. After the trams are placed, the other portions of the pavement are completed. After the surface has been well rammed with a wooden rammer, it is watered and covered with a bedding of sand ¾ inch deep, so as to fill the joints by degrees. On steep gradients, the surfaces of the trams are grooved diagonally.

Schedule of Prices in Milan.

Granite blocks for trams, from the quarries of St. Fedelino, including labour for setting, per lineal yard		10s. 3½d.
Granite blocks from Mont Orfano per lineal yard		8s.
Do. do. do. not including setting	„ „	6s. 10d.
Paving with cobbles	per square yard	10½d.
Pit sand	per cubic yard	1s. 6¼d.
River sand, from the Ticino or the Adda	„	3s. 0½d.
Screened pit gravel	„	1s. 1¾d.
Do. do. from the Adda	„	2s. 11d.
Cobbles for paving, from the Adda or the Brembo	„	5s. 8½d.
Do. do. from the Ticino	„	4s. 3d.
Do. do. from the Lambro or the Senese	„	4s. 10¼d.
Stone gratings for gullies	each	13s. 9d.
Wages:—Stone mason	per day	2s. to 2s. 9d.
Labourer	„	10½d. to 1s. 7d.
Pavior	„	10½d. to 1s. 7d.
Labour:—Dressing ends of granite blocks to form joints	per square yard	2s. to 3s. 4d.
Dressing sides of blocks	„	1s.
Dressing upper surface of blocks	„	2s. 6d.
Repairs:—Removal and rebedding of stones, including dressing upper surface and ends	per lineal yard	1s. 1d.
Removal, without re-dressing	„	10d.
Cutting diagonal grooves ⅜ inch deep, 4 inches apart, on trams	per square yard	8d.

From these prices, it is calculated that the cost of a line of tramway, comprising two lines of trams and the intermediate paving, amounts to £1,778 per mile.

On wide thoroughfares, as at Milan, there are three or four lines of granite tramway. But, on some lines, the interspace is paved with Dutch clinker bricks.

A Member of the Committee of the Society of Arts proposed a new form of tramway and road, in which a line of asphalte trams is laid on a road formed of hard concrete. See Fig. 55.

The design for a concrete road is based upon the prin-

Fig. 56.—Proposed Roadway with Wheel-tracks of Asphalte.

ciple of Mr. Mitchell's concrete macadam, already noticed. Another design by the same member is illustrated by Fig. 56.

CHAPTER XII.

WOOD PAVEMENT.

Wood-paving, originally, was, like granite-paving, laid in large blocks, and without any width of joint or interspace,—like the hexagonal paving introduced from Russia, measuring 8 or 9 inches across; and like Carey's original paving, in blocks 6 inches wide, and 12 inches long, a sample of which remains to this day in Bartholomew Lane. But, like granite-paving, wood-paving best answers its purpose when it is restricted, in the width of sets or blocks, to 3 inches, 3½ inches, or 4 inches, with interspaces, to afford the necessary stability for a minimum depth, and the necessary foothold for horses.

Wood-paving is, by almost universal practice, laid with the fibre in a vertical position,—on end. This is the best position for wear, for endurance, and for safety.

According to the best experience of wood-paving, it should consist of plain rectangular blocks, solidly set upon a foundation of concrete, with water-tight joints. A wood pavement so constructed as to fulfil these conditions, gives satisfaction on the five points of convenience, cleansing, maintenance, safety, and durability. Unless the foundation be rigid, it is impossible to preserve a sound pavement; and as in macadam, so in wood-pavement, the dogma of elastic action has been exploded by experience, for it was found that such a degree of elasticity as is afforded by the reaction of vertical wood-fibre, against a vertical pressure, is quite sufficient to absorb the shock of a horse's hoof, and to elude the strokes of loaded wheels.

As with granite sets, so with wood blocks, the gauge of a horse's hoof is the measure of the proper maximum width. The most common width of blocks is 3 inches; but they are sometimes made 3½ inches or 4 inches wide. The advantage of the narrower width consists in this, that, besides affording a more ready foothold, narrower blocks have more stability than wider blocks of the same depth; and that, in short, the narrower the block, the shallower it may be made,—so economising material. Mr. Carey's practice affords an illustration in point. His original blocks, which were 6 or 7 inches wide, were made 8 inches and 9 inches deep; according to his latest practice, his blocks are only 4 inches wide, and they are only 5 inches or 6 inches in depth. The course of the practice of wood pavement is strictly analogous to that of granite-set pavement. It may be added, that the length of a block should be suitably proportioned to the width. A length of 12 inches for a block 3 inches wide, has been tried and found to be too much:—the blocks were subject to splitting across.

The normal dimensions of wood blocks in current practice are:—width, 3 inches; depth, 6 inches; length, 9 inches. These are in the proportions of 1, 2, and 3.

The next essential provision is an interspace of suitable width between the courses of blocks, to make a good joint, affording, at the same time, foothold where the joints supply the only fulcrum. The blocks are set end to end in each course; and the width of interspace between the courses varies from about ¼ inch to ¾ inch. Experience has decided that the narrower the interspaces the better for the durability of the pavement.

The streets of the City of London afford the best and most exhaustive available experience of wood-pavements.

CHAPTER XIII.

CAREY'S WOOD PAVEMENT.

CAREY's pavement, already described, page 17, was the first durable pavement that was laid in the City. In July, 1841, this pavement, of pine blocks, was laid in Mincing Lane, and in October and November, 1842, it was laid in Gracechurch Street, covering an area of 1,623 square yards. The blocks were from 6¼ to 7¼ inches wide, 13 to 15 inches long, 9 inches deep for Mincing Lane, and 8 inches deep for Gracechurch Street. The foundation consisted of a layer of Thames ballast.

The cost for the wood pavement of Gracechurch Street, including a foundation of ballast, but not excavation, was 14s. 2d. per square yard. In 1844, two years after the pavement was laid, it was "much worn," and there were "several short holes in various places." It had been for a considerable time in a bad condition, when it was relaid in September, 1847, five years after it was laid; and a portion was relaid a second time in 1850, whilst ordinary repairs were made when required. In 1853, the paving was worn out, having been down 11 years, and it was replaced with new paving of the same kind, at a cost of 12s. 8d. per square yard. The second pavement was relaid in 1857, four years after it was laid, and again in 1861; and in June, 1865, it was removed, having been down for 11 years and 7 months. The costs of laying and maintaining the two successive pavements were as follows:—

TABLE No. 31.—CITY OF LONDON:—COST OF CAREY'S WOOD PAVEMENT IN GRACECHURCH STREET, 1842 to 1865.

Costs.	1st Pavement; duration 11 years.		2nd Pavement; duration 11·6 years.		Both Pavements; duration 22·6 years.
	per square yard.	per square yard per year.	per square yard.	per square yard per year.	per square yard per year.
	s. d.	s. d.	s. d.	s. d.	s. d.
First cost, including a foundation of ballast	14 2	1 3½	12 8	1 1	1 2¼
Relays and repairs	4 0	0 4½	17 1½	1 5¾	0 11¼
Total expenditure	18 2	1 8	29 9½	2 6¾	2 1½
Deduct value as old material	1 0	0 1	1 0	0 1	0 1
Net total cost	17 2	1 7	28 9½	2 5¾	2 0½
Add, cost for gravelling	..	..	..	..	0 1½
Total cost of the pavement for 22·6 years, 1842–65					2 2

The foregoing statement exhibits the cost for two completely new wood pavements laid and worn out in Gracechurch Street. They were succeeded by a third new pavement, on the same system, costing 11s. 6d. per square yard, laid in 1865. This was taken up, half worn, in 1871, having been laid down 6 years; and replaced by asphalte. The cost for maintenance during that time was 6s. 11d. per square yard, or 1s. 1¾d. per square yard per year. The cost for maintenance appears to have been a very variable item of expenditure. The costs for the three successive pavements are here brought together for comparison:—

Carey's Wood Pavement. Gracechurch Street.	Duration. years.	Cost for maintenance per square yard per year.
		s. d.
1st pavement	11	0 4½
2nd ditto	11·6	1 5¾
3rd ditto	6	1 1¾
Totals	28·6	0 11¾

The fluctuations of cost may not be readily accounted for; but it is clear that the generally increased traffic was one cause of the greater expenditure on the second and third pavements. The general total cost for maintenance may be taken at 1*s.* per square yard per year.

Carey's pavement was laid in Mincing Lane in July, 1841, and was taken up and replaced by pavement of the same kind, 9 inches deep, in August, 1860, having been down 19 years and 1 month. This was the oldest wood pavement then in existence in the Metropolis. During that period it was turned and relaid, and again relaid, the tops of the blocks having been cut off; and it was at other times extensively repaired. There is no other instance in this country of so long a duration in a wood pavement in a public thoroughfare. The second pavement, though not worn out, was taken up in August, 1873, to be replaced by asphalte. The following table, No. 32, shows the first costs and the costs for maintenance of Carey's wood pavement in Mincing Lane:—

TABLE No. 32.—CITY OF LONDON:—CAREY'S WOOD PAVEMENT IN MINCING LANE, 1841—1873.

Costs.	1st Pavement: duration 19·1 years.				2nd Pavement: duration 13 years.				Both Pavements; duration 32·1 years.	
	per square yard.		per square yard per year.		per square yard.		per square yard per year.		per square yard per year.	
	s.	d.	s.	d.	s.	d.	s.	d	s.	d.
First cost	14	4	0	9	9	2	0	8½	0	8¾
Relays and repairs	13	4	0	8·4	22	6¾	1	8¾	1	1½
Total expenditure	27	8	1	5·4	31	8¾	2	5¼	1	10¼
Deduct value as old material	1	0	0	0·6	1	0	0	1	0	0¾
Net total cost	26	8	1	4¾	30	8¼	2	4¼	1	9½

In Mincing Lane, as in Gracechurch Street, the cost for maintenance of the second pavement was much greater than that of the first pavement, and for the same reason; thus:—

Carey's Wood Pavement. Mincing Lane.	Duration. Years.	Cost for maintenance per square yard per year. s. d.
1st pavement	19·1	0 8·4
2nd ditto	13	1 8¾
Totals	32·1	1 1½

The two series of Carey-pavements, in Gracechurch Street and Mincing Lane, have supplied valuable experience of wood paving, and they have, therefore, been treated here somewhat in detail. The following table, No. 33, is given by Colonel Haywood, comprising the general results of the duration and cost of Carey's pavements in the City:—

TABLE No. 33.—CAREY'S WOOD PAVEMENTS IN THE CITY OF LONDON, CONSISTING OF LARGE BLOCKS:—DURATION AND COST.

Street.	Date when laid new.	Duration. Yrs. Mon.	First cost per square yard. s. d.	Cost for repairs per square yard. s. d.	Total average cost per square yard per year. s. d.
LARGE TRAFFIC.					
Cornhill	May, 1855	10 2	12 2	17 4¾	2 11
	July, 1865	6 8	11 6	8 9¾	3 0½
Gracechurch Street	Nov. 1853	11 7	12 8	17 1½	2 6¾
	June, 1865	6 0	11 6	6 11	3 0¼
Lombard Street	May, 1851	9 4	9 6	6 0	1 7¾
	Sep. 1860	10 7	9 2	20 2	2 9

TABLE No. 33—continued.

Streets.	Date when laid new.	Duration.		First cost per square yard.		Cost for repairs per square yard.		Total average cost per square yard per year.	
		Yrs.	Mon.	s.	d.	s.	d.	s.	d.
SMALL TRAFFIC. Lothbury {	May, 1854	12	3	12	6	28	4¾	3	4
	Aug. 1866	6	1	12	6	3	5¼	2	7½
Mincing Lane {	July, 1841	19	1	14	4	13	4	1	5¼
	Aug. 1860	13	0	9	2	22	6¾	2	5¼
Bartholomew Lane {	May, 1854	12	3	12	6	17	5¾	2	5¼
	Aug. 1866	5	5	12	6	3	11¼	3	0¼

Cost for Foundations included, but no Excavation. No deduction is made for the value of old material.

All the second wood pavements referred to in this table were removed before they were worn out, and replaced with asphalte, except the last. Nearly all of them would, by relay, with the insertion of some new wood, have endured a few years longer. Taking all the pavements together, the following averages are deduced:—

Averages for Carey's Pavement.	Duration of Pavements. Years.	Total cost per square yard per year.
Streets of large traffic	9·05	2s. 7½d.
Streets of small traffic	11·33	2s. 4½d.
Total averages	10·19	2s. 6d.

It may be added that Carey's paving in Bartholomew Lane has recently (February, 1877) been extensively renewed. It was newly laid in August, 1866, and the original paving, in its entirety, has thus lasted 10½ years.

As just now mentioned, Carey's recent wood pavement consists of wood blocks, 4 inches wide, and 5 inches or 6 inches deep, according to the traffic of the street, and

9 inches long. The ends only of the blocks are formed on Mr. Carey's original model, Fig. 11, with double-bevelled surfaces, salient and re-entering to the extent of 5-16 or ⅜ inch, which come together for the purpose of preventing the shifting of the blocks, and of distributing the pressure on one block over the contiguous blocks. The paving is placed on a bed of ballast or sand, 2 inches deep, laid on the old bed of the street; and the joints, ⅜ inch wide, are grouted with lime and sand. It is needless to remark that this pavement can only endure when it is laid on a previously existing foundation; otherwise, if laid with 2 inches of sand on a loose excavated bottom, it cannot be durable.

CHAPTER XIV.

IMPROVED WOOD PAVEMENT.

This pavement was introduced from the United States, where it has been known as the Nicolson pavement. It has already been noticed, page 14. The first piece was laid, in 1871, in the City of London; but the practice of the proprietors of this pavement has, since that time, been considerably modified. Their practice has been, after excavating to the profile and section of the street, to lay a bed of sand, fine mortar stuff, or dry earth, 4 inches deep, on the bottom. On this bed, two layers of 1-inch deal boards, previously dipped in boiling tar, were close-laid transversely and longitudinally. On these boards, wood blocks, 3 inches wide, 6 inches deep, and 9 inches long. also dipped in tar, were placed in transverse courses, the heading joints abutting, whilst the courses were interspaced by $\frac{3}{4}$-inch fillets of wood nailed to the floor, and through the blocks. The $\frac{3}{4}$-inch joints were made tight by a boiling mixture of tar and pitch poured into the spaces to a depth of an inch or two. The spaces were then filled up with dry ballast, which was rammed down by means of a flat caulking iron made for the purpose; and filled up with the tar mixture. The surface was then strewed lightly with small gravel, over which boiling tar was scattered from a pail fitted with a flat-nozzled spout. Finally a sprinkling of sand was thrown over the surface. The gravel and tar coating was worked into the surface by the wheels of traffic. Spaces, $\frac{1}{4}$ inch wide, were left between the heads of the blocks in each course, and grouted.

The tar and pitch mixture, in the proportion of 30 gallons of tar and 4 cwt. of pitch, were boiled together in an iron kettle, for at least four hours, until they acquired such a consistency that, when cold, the compound should become elastic and tough. When not sufficiently boiled, the mixture became brittle when it cooled, and was broken up by wheels and feet. The object of the elastic binding was to form a surface which should continue water-tight, as had long been effected by the same means in the granite pavements of Manchester.

The principal advantage claimed for this system was, that the flooring of planks formed an elastic foundation, and distributed the weight applied on one block over a large surface, whilst the additional elastic action minimised the wear of the blocks. There was an inconsistency in the principle of the construction, for it was presumed that the bed of sand would solidly support the flooring, at the same time that the flooring was expected to yield elastically under pressure. Again, so soon as leakage, even to the slightest extent, commenced, by which surface-water was allowed to penetrate downwards between the paving blocks, there was nothing to prevent its reaching to and saturating the substratum of sand; since the boards, although close-laid, were not tongued, and the water could pass through them with comparative freedom. The saturated substratum became mobile, and subject to movement under variations of pressure. Consequently, when a load passed over the surface, the boards, opposing an inconsiderable resistance to deflection, were pressed downwards by the load, into the quicksand, and they recovered their normal position when the load passed away. In this manner, a pumping action was set up, and the sand and water, mixed with other loose matter at the bottom, was pumped up to the surface in the form of mud and slime. Thus the pavement became gradually undermined, and the undermining process was accelerated by the form of the pavement itself, which presented a continuous diaphragm under which the exhausting process was extended, as by a diaphragm-pump. It is scarcely necessary to add, that the wetter the weather, the greater was the action of undermining.

But, in addition to the general liability to leakage of water through the pavement, there is a special difficulty in keeping it watertight at the kerbs, where it is comparatively overhung and unsupported, and where there is, at the same time, a constant supply of water for penetration, so long as there is any water in the channels.

A serious consequence of the flexibility of the pavement is the numerous breakages of the blocks by splitting, caused by the unequal stress and leverage of the load on blocks which are supported by a floor partly non-resisting and partly resisting.

A small sample of the Improved Wood Pavement, 468 square yards, was laid in Bartholomew Lane, in the City of London, in December, 1871. This is a short street of very little and light traffic. This pavement, which was the first of its kind laid in London, continues now (1877) in fair condition, but many of the joints are open.

The enduring power of the pavement has also been tested in thoroughfares of heavy traffic. In August, 1872, and January, 1873, an area of 6,066 square yards was laid, with 4-inch blocks, in King William Street and Adelaide Place, on the approach to London Bridge, from the Statue. In the course of the following three years, the foundation was rather extensively undermined, by the causes which have already been described,—particularly near the kerb on one side;—so much so as to cause the pavement to be deflected in many places to the extent of an inch under the wheels of passing vehicles, whilst the mud was seen bubbling up from below. The pavement was lifted about the month of January, 1876; and it was found that the substratum of sand had been partially washed away, and that, in the lines of regular traffic, the blocks had lost about $\frac{7}{8}$ inch in height, on an average, by vertical wear. The pavement was renewed on the same system.

The Improved Wood Pavement which was laid, in November, 1874, in Bishopsgate Street Without, in the City of London, is a more recent instance of the undermining influences set up under the action of heavy traffic, brought to a head during the past winter (1866-67), which has been a season of unusually heavy and continuous rain. The pavement, though it had not (in February, 1877) been down $2\frac{1}{2}$ years, had already been extensively undermined, and it was observed to be subject to deflection to the extent of half an inch, at some places, under passing wheels. At the same time, the collateral symptom of mud working up through the joints, became manifest.

The obvious remedy for the pumping action through the pavement is to lay a rigid foundation under the pavement. The flooring, to be stiff enough, would require to be constructed of two layers of 3-inch battens. But a less costly method of providing rigidity has been adopted in the construction of the new pavement now being laid to replace the defective pavement above referred to, which has been taken up. There is a thin bed of concrete, upon which a single layer of 1-inch boards is laid, and the blocks, $3\frac{1}{2}$ inches wide, are laid, as before, upon the boarding. The joints between the courses of the blocks are only $\frac{1}{10}$ inch wide; they are filled for half the depth with an artificial asphalte, consisting of chalk, pitch, and coal-tar, boiled together, and are filled to the top with a grouting of gravel and blue lias lime. It may be added that, for a width of 18 inches at each side of the roadway, at the channels, the boards are not laid, but their place is occupied by an additional depth of concrete, to afford a rigid support, so as to resist deflection, and the entrance of water at the edges.

The carriage-way of Ludgate Hill was paved on the original system of the Improved Wood Company in November, 1873. The pavement has been worn into bumps and hollows, and is about to be entirely renewed. It has been proposed to renew the pavement by laying the wood blocks on a foundation of concrete 6 inches thick, with a 2-inch bed of sand intervening.

CHAPTER XV.

OTHER WOOD PAVEMENTS.

Ligno-Mineral Pavement.—This pavement, known also as Trenaunay's system, was imported from France. It was the first system of wood pavement that was provided with a hard concrete foundation, and moulded exactly to the required curve of the wood. The concrete is laid with cement or blue lias lime, and besides supplying a firm foundation, it closes the surface against underground exhalations. Blocks of hard wood,—oak, elm, beech, or ash,—are laid on the concrete; their upper and lower surfaces are cut at an oblique angle, of about 60°, to the grain, in order that the fibre may be presented obliquely to the wearing surface, and that each block may be partially supported by the next block in the course upon which it leans. The courses are laid transversely, and the inclination of the blocks is alternately to the right and to the left in the successive courses. The upper edges of the blocks are chamfered. A groove is cut horizontally along the sides of the blocks, near to the base, which is filled with the asphaltic mastic used for jointing—so forming an additional tie. The joints are only partially filled with mastic, and they are filled up with a grouting of lime and sand.

The blocks are previously subjected to a process called mineralisation: they are placed in drying chambers heated with steam-coils, and exposed to a continuous current of air. When dried, the blocks are withdrawn, and plunged into tanks containing the mineral oils, consisting of hydrocarburets; they quickly absorb the oils, with which, it is said, the wood becomes saturated, and are so rendered tougher and more durable.

Though the employment of hard woods is a specialty of this process, mineralised firs also are used, and they are placed upright, with vertical fibres. But it is claimed that, in hard wood, a less depth of block suffices than in soft wood, and that thus the costs for the two classes of wood are equalised.

The first piece of this pavement laid in the City of London, was laid in Gracechurch Street, near Talbot Court, in August, 1872. On a bed of concrete, 4 inches thick, mineralised elm blocks were laid, bedded in Portland cement. The blocks were $3\frac{1}{2}$ inches wide, $7\frac{1}{2}$ inches long, and only $4\frac{1}{2}$ inches deep. The joints were partially filled with heated mastic, and filled up with a grouting of lime and sand. The pavement was relaid in 1875, after having lain three years, and a portion of it was replaced with new blocks.

The next piece of paving was laid, in 1874, in Fore Street, which is entirely paved on this system. On a concrete bed of blue lias lime, 6 inches thick, hard-wood blocks were placed, $3\frac{1}{2}$ inches wide, $7\frac{1}{2}$ inches long, and $4\frac{1}{2}$ inches deep. The wood used for nearly the whole of the surface was beech; supplemented by blocks of oak and of elm. The lines of channel are formed throughout of elm blocks.

Lastly, Coleman Street was, in 1875, paved with vertical blocks of mineralised fir, 3 inches wide, 9 inches long, and 6 inches deep, with $\frac{3}{8}$-inch joints; on a concrete foundation 6 inches deep.

Asphaltic Wood Pavement. — On this system, originally patented by Copland, a solid concrete foundation, 6 inches thick, is laid to the curvature of the road; it is composed of blue-lias lime and ballast, in the proportion of 1 to 5 or 6. Upon this foundation, a coat of mastic asphalte, $\frac{5}{8}$-inch or $\frac{3}{4}$-inch thick, is laid, as a bedding for the wood blocks. The blocks are 3 inches wide, 6 inches deep, and 9 inches long, of Baltic fir, laid in transverse courses, butt-jointed, with $\frac{10}{8}$-inch interspaces. The interspaces or joints are run up with melted asphalte to a depth of about 1½-inches. The correct interspacing of the courses is ensured by placing long strips of wood $\frac{10}{8}$-inch thick, against each course as it is laid. The asphalte, when poured, remains for two or three minutes in a fluid condition, and it partially re-melts the coating on which it rests, and unites with it, whilst it adheres firmly to the blocks. The whole structure thus becomes solidified as one mass, and the joints are filled up with a grouting of sand and hydraulic lime. The grouting serves as a non-conductor of heat, and fixes the grit and gravel which it is usual to strew over a newly-laid wood paving, for the purpose of indurating the surface of the wood.

The first piece of Copland's pavement in London was laid at the east end of Cannon Street, in 1874, on a concrete foundation, 9 inches thick, made with Portland cement; a layer of mastic asphalte, $\frac{3}{4}$-inch thick, was spread over the concrete, and carried the wooden blocks, which were of soft wood, and had been steeped in tar. The blocks were laid in courses across the street, with $\frac{5}{8}$-inch joints between the courses. Each block had two holes $1\frac{1}{8}$-inch in diameter, and $\frac{5}{8}$-inch deep, bored in each side; and when the joints were filled to half their depth with "liquid asphalte," the asphalte of course occupied the holes. The upper part of the joint was then rammed with screened gravel, and grouted with tar and asphalte, and the pavement covered with screened gravel and sand.

The lateral holes have not, in later pavements on this system, been made in the blocks. It was found, no doubt, that they were an entirely useless refinement.

OTHER WOOD PAVEMENTS.

The asphaltic wood pavement appears to have answered well. In Bristol, it has been laid for upwards of two years, in Broad Street and at the Exchange, and Mr. F. Ashmead, the Engineer, reports that it is wearing well. He maintains that the principal advantage of the pavement is "its imperviousness to water." It is stated that, on lifting a portion of this pavement after having been down two years, the foundation was found to be as dry as when the pavement was laid.

Harrison's Wood Pavement.—This pavement is analogous to the Asphaltic Wood Pavement. Upon a concrete foundation, it is proposed to place strips of wood, 2 inches wide by ½-inch in thickness. Upon these, 3-inch wood blocks are to be placed, and heated asphalte is to be poured into the joints, penetrating under and adhering firmly to the blocks. On this system, an under coating of asphalte would be formed in sections by the successive pourings into the joints, instead of in large sheets as formed by the Asphaltic Wood Company. This system does not recommend itself. The multiplication of attachments by pieces of wood is objectionable.

Henson's Wood Pavement.—On a solid substratum of blue lias lime concrete, 6 inches thick, covered by a 2-inch layer of cement-concrete of a finer quality, a coat of ordinary roofing felt is spread;—the felt having been previously saturated with a hot asphaltic composition of distilled tar and mineral pitch. On this felt, as on a carpet, cushiony and impervious to moisture, blocks of Swedish yellow deal, containing resin sufficient for preservation, are laid. The blocks are 3 inches wide, 6 inches deep, and 9 inches long, are placed with the grain upright, and laid closely together end to end, in rows across the street. The rows are also driven together and close-jointed with a strip of saturated felt in each joint. The width of the interspaces is thus reduced to the simple thickness of the felting, and does not exceed, if it even amounts to, a quarter of an inch. At intervals of every three or four rows, a row of blocks grooved along the middle is laid, to aid in giving foothold. The surface is dressed with a hot bituminous compound and fine clean grit.

This pavement was laid in Oxford Street in two lengths, between Princes Street and Marylebone Lane, and between Hereford Gardens and Edgware Road; and was opened for traffic in December, 1875. For the 18 months (to June, 1877) during which this pavement has been open, it has, according to the report of Mr. H. T. Tomkins, the Surveyor, given very great satisfaction, and it continues in excellent condition. A piece of Henson's wood pavement was laid in Leadenhall Street, in the City of London, in August, 1876.

Norton's Wood Pavement.—This pavement was evidently designed as a compromise, to do away with a foundation of concrete, whilst securing the advantage of a firm base. A thin bed of ballast is laid on the subsoil, and brought to the profile of the street, to receive the pavement. The pavement consists of slabs or structures, 7 feet long, and 3 feet wide, composed of blocks of yellow Baltic timber, 3 inches wide, 6 inches deep, and 7 inches long, fixed to a backing of 2-inch boards, with a strong bituminous cement. The joints are run with the same material. Between the slabs, the joints, ¼-inch wide, are filled with powdered rock asphalte, in a heated state, well rammed and levelled to the surface of the blocks. A small piece of paving on this system was laid at the east end of Cannon Street, in 1874. It has been occasionally under repair, and is now (February 1877) in bad condition :—bumpy, and depressed at the edges of the slabs.

Mowlem's Wood Pavement.—This pavement reverts to the simple type of wood pavement, formed of fir blocks, inches wide, 6 or 7 inches deep, and 9 inches long, with joints, transversely, ¼-inch wide, filled with a grouting of blue lias lime and sand. It is laid on a foundation of concrete made with Portland cement, 6 inches thick. The earliest piece of Mowlem's paving was laid in Duke Street, Smithfield, in June, 1873. A small piece was laid at the east end of Cannon Street, in September, 1873. Portions of the blocks were creosoted.

Stone's Wood Pavement.—A concrete foundation is formed, compressed by machinery. Grooves, ¾-inch deep, and 3 inches apart, are cut by machinery in the surface of the concrete. Wood blocks, 3 inches wide, 5 inches deep, and 6 inches long, are placed on the concrete, being shaped to fit into the grooves. The blocks are in transverse courses, with 1¼-inch joints, which are filled with gravel, and run with heated tar. A specimen was laid in King William Street, near to Clement's Lane, in July, 1873. Having been completely worn out by the traffic, it was taken up and replaced by the Improved Wood Pavement, in October, 1876. This new pavement was constructed with one layer of boarding on concrete and sand.

Gabriel's Wood Pavement.—Wood blocks, 3 inches wide, 6 inches deep, and from 7 to 11 inches long, are placed, with the intervention of a little sand, on a foundation of concrete, composed of Thames ballast and Portland cement or ground lias lime, of from 6 to 9 inches in thickness. Two small fillets are secured to one side of each block, so as to keep the blocks steady during the process of grouting. The joints are filled with a grout of lime and sand, and the whole surface of the pavement is covered with a layer of hoggin and sand. The eastern and the greater part of the south side of St. Paul's Churchyard were laid with this pavement in 1875.

Wilson's Wood Pavement.—This pavement is formed of blocks 3 inches wide, 6 inches deep, 8 inches long, laid on a foundation of blue lias concrete 6 inches deep, on which a ¼-inch layer of sand is spread, to bed the blocks. Two small fillets are secured by brads on the side of each block, so as to keep the blocks steady until the process of grouting is completed. The grouting for the joints consists of blue lias lime and sand. A piece of this pavement, 419 square yards in extent, was laid in Fetter Lane, in October, 1876.

TABLE No. 34.—CITY OF LONDON:—WOOD PAVEMENTS.

Situation.	Name of Wood Pavement.	Length.	Area.	Date when the paving was completed.
		Yards.	Sq. yds.	
Bartholomew Lane	Carey's Wood	40	468	Jan. 1872
Birchin Lane	,,	28	77	June, 1866
Jewry Street	,,	42	253	Feb. 1872
Little George Street	,,	24	148	Feb. 1872
Cannon Street	,,	—	9,051	Sept. 1874
Houndsditch, N.W. half	,,	—	1,826	May, 1874
Barbican, a third part	,,	—	800	Oct. 1875
Bartholomew Lane	Improved Wood	48	392	Dec. 1871
Great Tower Street and Seething Lane	,,	76	448	Aug. 1873
King William St., between Gracechurch Street and Cannon Street, and Adelaide Place	,,	311	6,066	Jan. 1876
Ludgate Hill	,,	266	2,639	Nov. 1873
Aldersgate Street, &c.	,,	—	6,884	Sept. 1874
Barbican, a third part	,,	—	606	Nov. 1875
St. Mary Axe, &c.	,,	—	1,662	July, 1875
Bishopsgate Street Without	,,	—	—	Feb. 1877
Gracechurch Street, near Talbot Court	Ligno-Mineral	27	440	Aug. 1872
Fore Street	,,	—	3,822	Dec. 1874
Coleman Street	,,	—	2,291	June, 1875

OTHER WOOD PAVEMENTS.

TABLE No. 34—*continued.*

Situation.	Name of Wood Pavement.	Length.	Area.	Date when the paving was completed.
		Yards.	Sq. yds.	
Duke St., Smithfield	Mowlem's Wood	134	676	June, 1873
Houndsditch, S.E. half	,,	—	1,832	May, 1874
Wormwood St., &c.	,,	—	689	June, 1875
Cannon Street, sample at east end	Asphaltic Wood	—	299	July, 1874
Barbican, a third part	,,	—	785	Oct. 1875
St. Bride Street	,,	—	1,774	Nov. 1876
Queen Street	,,	—	2,169	July, 1876
Cannon Street, sample at east end	Norton's Wood	—	—	1874
St. Paul's Churchyard, east end and part of south side	Gabriel's Wood	—	4,907	Jan. 1876

In the Appendix is given a table by Colonel Haywood, showing the condition of the wood pavements in the City of London, as existing on the 1st February, 1877.

CHAPTER XVI.

COST AND WEAR OF WOOD PAVEMENTS.

Cost.—The first cost of wood pavements in the City of London, including foundation, but not excavation, has been as follows:—

	Laid.	First cost.
IMPROVED WOOD.		
King William Street	1873	18s. 0d. per square yard.
Ludgate Hill	1873	18s. 0d. ,, ,,
Great Tower Street and Seething Lane	1873	16s. 0d. ,, ,,
Bartholomew Lane	1871	16s. 0d. ,, ,,
CAREY'S WOOD.		
Bartholomew Lane	1872	12s. 6d. ,, ,,
Houndsditch	1874	13s. 6d. ,, ,,
MOWLEM'S WOOD.		
Duke Street	1873	15s. 3d. ,, ,,
Houndsditch	1874	17s. 0d. ,, ,,

The Asphaltic Wood Pavement has been laid in Bristol for 14s. 6d. per square yard.

With regard to the cost for maintenance of wood paving, the Improved Wood Company contracted to maintain the pavements of King William Street and of Ludgate Hill, 1 year free, and 15 years at 1s. 6d. per square yard per year; and the pavements of Great Tower Street and Seething Lane, 1 year free, and 15 years at 1s. 3d. per square yard. These terms are equivalent to a total cost per square yard per year for 16 years, of 2s. 6¼d. for the first and second streets, and 2s. 2d. for the third. But in King William Street, the pavement was renewed three years after it was first laid; so that the actual total cost must be something considerably greater than the contract cost.

The Asphaltic Wood Paving Company maintain the Bristol pavement for 1s. per square yard per year for 10 years; equivalent to a total cost for this period of 2s. 5¼d. per square yard per year.

Messrs. Mowlem and Co. maintain the pavement of Duke Street for 2 years free, and 3 years at 1s. per square yard per year. Also, the Houndsditch pavement, for 2 years free, and 5 years at 9d. per square yard per year.

Mr. Carey maintains the Houndsditch pavement, for 2 years free, and 5 years at 1s. per square yard per year.

Mr. G. J. Crosbie Dawson* gives the cost of the Ligno-Mineral Pavement as follows:—

	Per square yard.
Hard wood, including concrete foundation	14s. 6d.
Maintenance for 10 years, after the first 2 years	0s. 8d.
Yellow deal, including concrete foundation	11s. 6d.
Maintenance for 10 years, after the first 2 years	0s. 10d.
Footways, including foundations	5s. 0d.

The price of Mr. Henson's pavement, as laid in Oxford Street, is 12s. per square yard, if only 2 inches of Portland cement concrete is required; and 14s. if with 6 inches of the same concrete.

Mr. Ellice-Clark gave, in 1876, the cost for various pavements, if laid in Derby, from which the following particulars are deduced:—

Asphaltes.—Derby.	First cost per sq. yard.	Period of maintenance.	Cost per year.	Total in annual cost per year.
Limmer	16s. 0d.	13 yrs.	0s. 7d.	1s. 9¾d.
Val de Travers, with iron studs	17s. 6d.	—	—	—

* "Street Pavements," in the *Journal of the Liverpool Polytechnic Society*. 1876.

WOOD.				
Carey	14s. 9d.	10 yrs.	1s. 5d.	2s. 10½d.
Asphaltic Wood . .	14s. 3d.	13 ,,	0s. 7d.	1s. 8¼d.
Ligno-Mineral:—				
Yellow deal . .	11s. 6d.	17 yrs.	1s. 1½d.	1s. 9¾d.
Hard wood . .	14s. 6d.	17 ,,	0s. 5¼d.	1s. 3½d.
Improved wood . .	15s. 0d.	16 ,,	0s. 9¼d.	1s. 8½d.
GRANITE . . .	13s. 0d.	28 ,,	0s. 3⅛d.	0s. 8¾d.

In these estimates for Derby, except those for Carey's Wood, and the Improved Wood, the cost for a foundation of concrete is included, but no excavation.

That wood pavement, properly seasoned and properly laid, with watertight joints, whether lime-grouted or asphalted, may last for many years without suffering decomposition, has been amply demonstrated by the experience of Carey's pavements in the City of London, which were laid in 1841. In New York, it appears, the average duration of wood pavement is only four years, for "it is estimated that the wood pavements become worn out and useless at the rate of 25 per cent. per annum." The failure arises, not from simple wear, but from decay; and "the process of disintegration is much accelerated by the constant wear and pressure from the wheels of vehicles, and by the sudden changes of temperature which characterise our climate."* But the information is not of that precise character which would make it of use for comparison in England. It has, indeed, been averred that the wood was unsound when it was laid.

Wear.—In the City of London, the abrasive wear of wood pavement in the lines of traffic has been recently ascertained, in a few instances, by direct measurement, by the writer and others, to be approximately as follows:—

---

* Communications from the municipal authorities of New York and Paris on the "Results of their Experience of Wood Pavements in those Cities." 1876.

	WOOD PAVEMENT.	Time down.	Vertical Wear. Total.	Vertical Wear. Per year.
1. Adelaide Place, London Bridge	Improved Wood Co.	3 yrs.	⅞ inch	·29 inch
2. Bishopsgate St. Without	do. do.	2½ „	¾ „	·30 „
3. Tower Hill	Carey's	5 „	3½ „	·70 „
4. Bartholomew Lane	do.	5 „	¾ „	·15 „
5. Cannon St.	do.	2½ „	¾ „	·30 „

The amounts of wear in the first, second, and fifth instances are practically the same, namely, ·30 inch per year. The widths and the daily traffics at the places where the measurements were made, may be taken, for the first instance, at 41 feet and 16,000 vehicles in 12 hours; and for the second instance, 22 feet and 8,000 vehicles. For the fifth instance, in Cannon Street, the measurement was made just opposite Mansion-House Station, where there is much cross traffic. The normal traffic is upwards of 6,000 vehicles, and it may be taken as double, or 12,000 vehicles, at this spot, with a width of 36 feet. The comparison, then, stands thus—

	Width of street.	Traffic per day.		
1. Adelaide Place	41 ft.	16,000 vehicles,	or 390 vehicles per ft. wide	
2. Bishopsgate St. Without	22 ft.	8,000 „	or 364 „	„
5. Cannon Street	36 ft.	12,000 „	or 333 „	„
			362 „	„

From these data, it appears that, for a traffic of 362 vehicles in 12 hours per foot of width, the vertical wear amounts to ·300 inch per year. The wear for 100 vehicles per day per foot of width, is, therefore, equal to ($·300 \times \frac{100}{362} = $) ·083 inch, or $\frac{1}{12}$ inch per year.

Experience is wanting,—at least in England,—for determining the minimum depth to which wood-block paving should be worn, before being removed. It is said that, in some instances, ligno-mineral and other wood blocks have been worn down to a depth of 2½ inches when they were removed. Sufficient depth of block is required for stability, in wood-paving as in stone-paving; and, reasoning by comparison, it would appear that there may be some mistake in the first design of wood-paving, by limiting the depth, when new, to 6 inches. The motive for this limitation of depth, it can hardly be questioned, is one of economy in first cost; but when it is observed that Carey's earlier wood-pavements in the City of London, which had 8 inches and 9 inches of depth, have far surpassed in duration more recent pavements having only 6 inches of depth originally, it would be well for wood-paviors to try back, and revise their designs on the basis of a 9-inch depth of block for heavy traffic. In forecasting the amount of vertical wear to which such blocks may judiciously be subjected, it may be premised that, for equal original dimensions, the amount of vertical wear may be at least as much for wood, as it may probably be, and as has been indicated, for stone-sets; for the geometrical form of the wood-block admits of a more secure and more permanent joint being made than for the roughly squared stone-set. Empirically, then, the total vertical wear may be taken at 1 inch on a depth of 6 inches, and 4 inches on a depth of 9 inches, leaving stumps of 5 inches to be removed. But, it is scarcely necessary to add, that in order that a wood-pavement should be capable of undergoing such amounts of wear, it must be solidly and rigidly constructed with thoroughly seasoned close-grained wood and watertight joints, and thoroughly drained. Moreover, time, the destroyer, is an element, and, in making estimates, it is necessary to impose a limit to the duration of wood, independent of the resistance to wear. In the following table, the limit of endurance is taken at 20 years:—

TABLE No. 35.—WOOD PAVEMENT:—ESTIMATED DURATION.

Vertical wear, $\frac{1}{12}$ inch per year per 100 vehicles per day per foot of width.

Traffic per foot of width per year.	Height of blocks.			
	6 inches.	7 inches.	8 inches.	9 inches.
	Maximum amount of vertical wear.			
	1 inch.	2 inches.	3 inches.	4 inches.
Vehicles.	Years.	Years.	Years.	Years.
100	12	20	—	—
200	6	12	18	20
300	4	8	12	16
400	3	6	9	12
500	$2\frac{1}{2}$	5	7	$9\frac{1}{2}$

CHAPTER XVII.

ASPHALTE PAVEMENTS.

THE employment of asphalte for the carriage-way of streets was commenced in Paris, where, in 1854, the Rue Bergère was laid with Val de Travers asphalte. In 1858, three sides of the Palais Royal were laid with the material, which was brought to the ground in the state of rock crushed into small pieces, and was heated and powdered by decrepitators. The foundation of concrete was about 6 inches thick; the asphalte was 2·4 inches thick in the Rue St. Honoré and the Rue Richelieu, and 2 inches in the Rue de Valois. The price was 20 francs per square metre, or 13s. 4d. per square yard. The success of these pavements was, it is said, complete, in spite of an unpropitious season; and, in 1859, the Rue des Petits Champs was laid. But, the conditions were unfortunate. The street had been occupied by drainage works, and had been filled with loose soil to a depth of from 10 to 13 feet. On a 4-inch bed of concrete, 2 inches of asphalte was laid; the work was carried on in rain and snow, and in January the street was opened for traffic. Cracks in the roadway immediately appeared, owing to the constant wet during the process of laying, and the settling of the soil. Each damaged part was repaired in succession. The price paid for the work was 15 francs per square metre, or 10s. ½d. per square yard, including the concrete at 2 francs per metre, or 1s. 4d. per yard. This price was to serve as a basis for future proposals for re-paving streets with compound asphalte:—to the standard of 4 inches of concrete, and 2 inches of asphalte. The conversion of the street-pavements into asphalte work, on a large scale, was commenced in 1867. Seyssel-rock asphalte was laid in the Rue Richelieu and elsewhere, and, according to report, answered perfectly well.

Asphalte Pavements in the City of London.

The first pavement in asphalte, laid in the City of London, was laid of compressed Val de Travers asphalte, in Threadneedle Street, near Finch Lane, in May, 1869. The next pavement was laid, in October, 1870, in Cheapside and the Poultry. Asphaltic pavement was rapidly extended to other streets in the City, until, at the end of 1873, the total area of carriage-way covered by asphalte amounted to about 61,000 square yards, extending over 7,484 lineal yards, or $4\frac{1}{4}$ miles, of thoroughfare. From these data, it appears that the average width of carriage-way so laid was 24 feet 4 inches. The details are given in a following table. Meantime, the various kinds of asphaltic pavements that have been laid and tried in the City of London, are now to be described. The particulars for description have been gleaned, for the most part, from Colonel Haywood's reports.

Val de Travers Compressed Asphalte Pavement.—Carriage-ways are constructed with a foundation of concrete, of from 6 to 9 inches in thickness, according to the traffic in the street. The rock in its natural state is broken up and reduced to powder by exposure to heat in revolving ovens. It is then lodged in iron carts with close-fitting covers, and brought upon the ground, taken out, laid over the surface, and, whilst hot, compressed with heated irons into a homogeneous mass without joints. The finished thickness is from 2 to $2\frac{1}{4}$ inches, according to the traffic, and the material is further compressed and consolidated by the action of the traffic: by as much as 20 or 25 per cent., according to the statements of the Company.

The pavement laid in Threadneedle Street, already noticed, was laid on a foundation of concrete 8 inches thick; a coating of mastic, ½-inch thick, was laid over the concrete, and over this, 2 inches of Val de Travers compressed asphalte. The mastic was applied to expedite the work; but it has not been used in any other pavement formed in the City.

The next pavement that was laid,—in Cheapside and the Poultry,—was 2¼ inches thick, with a foundation of concrete, 9 inches thick. The cost was 1s. 9d. per square yard for concrete, and 16s. 3d. for asphalte: together, 18s. per square yard.

Val de Travers Mastic Asphalte Pavement.—In laying this pavement, the rock is ground to a fine powder, and placed in caldrons on the ground, with an addition of from 5 to 7 per cent. of bitumen. When heated, it becomes semi-fluid; then, 60 per cent. of grit or dry shingle is added to the melted mass, and after having been thoroughly mixed, the compound is spread over the bed of concrete in one layer. In March, 1871, George Yard, Lombard Street, was paved with this mastic asphalte, in a layer 1½ inches thick, on a bed of concrete 6 inches thick. The price, including concrete, was 12s. per square yard.

Limmer Mastic Asphalte Pavement.—The rock is broken up and mixed with clean grit or sand, of different sizes, according to the place in which the pavement is to be laid. A small quantity of bitumen is added to the materials. The mixture is heated and made liquid in caldrons on the spot, and the compound is run over the surface of the foundation of concrete, and smoothed with irons to the required profile. It is run in two layers, the lower of which is made with grit of a larger size than that of the upper layer. The total thickness of asphalte, when finished,

ASPHALTE PAVEMENTS.

is from 1½ to 2 inches, according to the traffic. This asphalte was first laid in Lombard Street, and the pavement was finished in May, 1871. The concrete was 9 inches, and the asphalte 2 inches thick. The price of the concrete was 2s. 8d., and of the asphalte 13s. 4d.; together, 16s. per square yard.

Barnett's Liquid Iron Asphalte Pavement.—This is made either of natural or artificial asphalte, mixed with pulverised iron ore or sesqui-oxide of iron, and a small proportion of mineral tar. The materials are heated in a caldron, and melted, then run over the surface, and smoothed like other liquid asphaltes. The usual thickness is 2¼ inches. It was first laid in Moorgate Street, in October, 1871. The price was 13s. 6d. per square yard, exclusive of foundation.

Trinidad Asphalte Pavement.—This is a compound of Trinidad pitch, broken stone, chalk, and other ingredients comminuted and mixed together. It is laid as a heated powder. A piece of this pavement was laid in Princes Street, in June, 1872, for which the price was 10s. 8d. per square yard. This pavement showed signs of wear directly it was opened for traffic. It was continually under repair, but it got cut to pieces, and was taken up in the end of October, and replaced by Val de Travers Asphalte.

Patent British Asphalte Pavement.—This pavement does not contain any natural asphalte. It is composed of certain oils, caustic lime, pitch, sawdust, and iron slag or grit ground to a powder. The mixture is heated in a boiler, and, in a semi-liquid state, run over the surface of a bed of concrete. This pavement was laid next the Trinidad asphalte, in Princes Street, in July, 1872, at the price of 12s. per square yard. By the month of December, it had suffered severely from the wear of the traffic, by which it was pulverised; and, after having been repaired, ineffectually, it was taken up in the same month and replaced by the Val de Travers Asphalte.

Montrotier Compressed Asphalte Pavement.—This is a natural asphalte from the mines of Montrotier, in the department of Haute Savoie, France. It was laid as heated powder, in Princes Street, at the north side, and compressed with rammers in the same manner as the Val de Travers asphalte. The work was completed in August, 1872. Price, 15s. per square yard.

The compressed asphalte of the *Société Française des Asphalte* is brought from the mines of Garde Bois in Seyssel. It was laid 2⅜ inches in thickness, in Princes Street, in July, 1872, on 9 inches of concrete. The price was 14s. 9d. per square yard.

Maestu Compound Asphalte.—This asphalte is brought from mines at Maestu, in Spain. The pavement is formed of blocks of asphalte compressed into the shape of bricks, about 6¼ inches wide, 2¾ inches thick, and 13 inches long, laid on a bed of concrete, and the joints grouted with bitumen. This pavement was laid in Threadneedle Street, in December, 1871, at the price of 12s. per square yard. It failed at places, owing to defective concrete; and, after having been repaired, it was taken up in January, 1872, 1½ month after it was laid.

Stone's Slipless Asphalte.—A foundation, 9 inches thick, is composed of a peculiar cement, compressed while in a fluid state by machinery, under a pressure of 112lbs. per square inch. A layer of asphalte, 2¼ inches thick, in a heated state, was laid on this foundation; it was composed of tar, cement, sand, and lead-ore, subjected to compression as the concrete was. A specimen of the pavement was laid in King William Street, near Nicholas Lane, in July, 1873; but it failed to sustain the traffic, and was taken up in September, after having been down only 5 or 6 weeks.

Bennett's Foothold Metallic Asphalte.—This pavement consisted of blocks about 2 feet square and 4 inches thick.

The lower portion—3 inches thick—of the blocks is composed of British manufactured asphalte and burnt ballast. The superstratum of the blocks, 1 inch thick, is composed of foreign asphalte, bone-dust, sulphate of lime, and certain metallic substances, including lead. The strata were fused together. Portions of the superstratum, 6 inches in diameter, at 8¼-inch centres, were raised above the general level of the surface, and made hard by some process, with the object of increasing the wearing powers of the pavement, and to preserve its convexity. A specimen of the pavement was laid in King William Street, near Nicholas Lane, in October, 1873; but, failing to sustain the traffic, it was taken up in January, 1874, after having been down three months.

Lillie's Composite Pavement.—This pavement was composed of asphalte, wood, and broken granite. A specimen for trial was laid in Lombard Street.

McDonnell's Adamantean Concrete Pavement.—This material was composed of blocks made of broken stone, chalk, lime, and clay, mixed with vegetable or mineral pitch or tar. A portion of the carriage-way of Carter Lane was, in April, 1869, paved with this material. The blocks were 18 inches long, 12 inches wide, and 6 inches deep; they were laid with ¾-inch joints, upon a solid foundation, and run up with an asphaltic composition. The price was 20s. per square yard. The pavement showed a serious degree of wear at the end of the first year after it had been laid; and at the end of 18 months it had to be extensively repaired by the contractor, according to the terms of his contract. In 1872, it was in such a bad condition, that, not having been repaired by the contractor, it was taken up in September of the same year, after having been down 3 years and 5 months. The traffic of the Lane did not exceed 700 vehicles in 24 hours.

Granite Pavements with Asphalte Joints.—Trials were made

of asphalte jointing for granite sets, as in the pavements of Manchester and other northern towns. A portion of Duke Street, Smithfield, was paved with Carnarvon granite sets, 3 inches wide, 6 inches deep, and 6 inches long. They were laid with 1-inch joints, which were filled with small clean pebbles, and run up with a composition of pitch and oil boiled together. The work was completed in May, 1868.

A piece of Aberdeen granite pavement was laid for comparison at the same time, contiguously, in the same street. The sets were 3 inches wide, and 7 inches deep, and were jointed with lime-grout, in the usual manner.

In 1871, the two pavements were in nearly the same condition; but the lime-grouted pavement showed, if anything, the greater amount of wear:—the difference having been ascribed by Colonel Haywood,—no doubt correctly,—to the comparative softness of the Aberdeen sets. Both pavements were removed in May, 1873, after a five-years' trial, and replaced by wood.

Another piece of this kind of paving, formed with Aberdeen granite sets, 3 inches wide and 9 inches deep, was laid in St. Paul's Churchyard, at the east end, in April, 1870, and removed in December, 1875, after having been down 5 years and 8 months. During this time, no repairs were done to the pavement. For the first three years, the surface continued in very good condition; afterwards, it gradually fell into disrepair, and, in the last year, it was in a very bad state. Small repairs were not done to it, in the way customary with lime-grouted sets, by reason of the difficulty and expense of taking up and relaying small detached portions:—which required a boiler and special apparatus to be brought to the ground.

Colonel Haywood reported that all the asphalte-jointed granite pavements were noisier and less pleasant to travel over than the pavements grouted with lime.

The following table, No. 36, gives the length and area of carriage-ways paved with asphalte, at December, 1873. Since this date, several additional pavements in asphalte have been laid:—

TABLE No. 36.—CITY OF LONDON:—EXTENT OF ASPHALTE CARRIAGE-WAY PAVEMENTS, AT DECEMBER 31, 1873.

(COLONEL HAYWOOD.)

Pavement.	Number of streets, &c., paved.	Length.	Area.
		Yards.	Square yds.
Val de Travers, compressed	20	4,185	34,876
Val de Travers, mastic	1	69	232
Limmer, mastic	8	1,446	8,477
Barnett's, mastic	6	1,705	16,544
Société Française, compressed (Soyssel)	1	39	327
Montrotier, compressed	1	40	346
Total number of pavements	37	7,484	60,802

NOTE TO TABLE.—The Val de Travers compressed pavements and the Limmer-mastic pavements for carriage-ways, comprised in this table, were, with three exceptions, made with 2 inches of asphalte; they were laid on concrete of from 6 to 9 inches in thickness. The exceptional thicknesses were 2¼ inches for Cheapside and the Poultry, as before noted; and 2⅛ inches for Gracechurch Street and Queen Street.

The pavements of Barnett's Iron Asphalte were laid on 9 inches of concrete, and were 2¼ inches thick. The pavement of the Société Française was 2⅝ inches thick, on 9 inches of concrete; and that of Montrótier was 2 inches thick.

From the experimental results just recorded, it is apparent that no preparation of asphalte compounds can compete successfully with simple asphalte as a material for roadways. Experience has amply confirmed Colonel Haywood's belief, expressed in his report of July, 1871:—
"It has been seen," he says, referring to the Val de

Travers compressed asphalte and liquid asphalte, the Limmer mastic, and Barnett's iron-asphalte mastic, "that the modes of compounding and forming the asphalte pavements differ materially, the one being a mineral unmixed and laid in a state of dry heated powder, the other three being composed of asphalte largely mixed with grit, sand, and other ingredients, and laid in a heated liquid state. There is consequently a marked difference between their structures, and I incline to the belief that the asphaltes which will form the most durable pavements for carriageways are those capable of being laid and compressed in the shape of heated powder."

In a report by Colonel Haywood, dated April 18, 1873, he described the condition of the pavements laid in the City of London, at the 1st March of that year, from which the following table, No. 37, has been compiled. Portions of some of the pavements were cut out under his direction, in order to ascertain whether they had lost materially in thickness. "There is no doubt," he says, "that the asphaltes have somewhat diminished in thickness under the wear of the traffic, but owing to inequalities when they were first laid, and to the compression which takes place in all those laid in the shape of heated powder, it would be difficult to ascertain the exact wear without making a very large number of openings."

By "holes" or "short holes," in the table, it was not meant that the asphaltes were worn down to the concrete foundation. The larger holes upon compressed asphalte are in most cases depressions caused by the traffic, and do not necessarily indicate surface wear. In the mastic asphaltes, on the contrary, they are for the most part the result of disintegration and wear of the surface.

Minute holes are noticeable in compressed asphaltes shortly after they are laid, which seem, after a time, to close up or disappear, whilst others open. The cause of this "flow of solids," which Colonel Haywood thinks may be due to the presence of foreign substances, or of moisture from the foundation, has not yet been satisfactorily explained.

The lengths of time during which these asphaltes had been down, at the date of the inspection, were 3 years and 9 months for one, 2 years and 2 months for two, less than 2 years for eighteen, less than 1 year for four, and less than 6 months for five.

TABLE No. 37.—CITY OF LONDON:—DURATION

VAL DE TRAVERS

Locality.	Approximate average width of pavement.	Vehicles in 12 hours in 1872-73.		Time down at March 1, 1873.
		Total.	Per foot of width	
	Feet.	Vehicles.	Vehicles.	Yrs. Mos.
Poultry: next kerb, general surface	22	7,997	363	2 2¾
Cheapside: next kerb, general surface.	30	8,949	298	2 2¼
Old Broad Street	24	2,473	103	1 11¼
New Broad Street	24	1,515	63	
Throgmorton Street	14	661	47	1 11
Milk Street	14½	516	36	1 11¼
Russia Row	7	97	14	1 11¼
Queen Street	15	2,292	153	1 10
Old Bailey	22	2,903	132	1 10
Gracechurch Street	21	4,730	225	1 7¼
Finsbury Pavement and Moorgate Street.	42	5,361	128	1 6¼
Moorgate Street	32	6,000	188	1 6¼
Wood Street	14½	980	68	1 5¼
London Wall	24	2,500	104	1 4¼
Threadneedle Street, near Finch Lane.	26	3,696	142	3 9
Ditto, east end	14	,,	264	1 2¼
Ditto, central	,,	,,	264	1 1¼
Mansell Street	—	498	—	1 4
Mansion House Street	46	13,767	300	0 8¼
Princes Street	24	5,628	234	0 3¼
Ditto	,,	,,	234	0 2¼

VAL DE

| George Yard | — | 58 | — | 1 11 |

ASPHALTE PAVEMENTS. 253

AND REPAIR OF ASPHALTE PAVEMENTS, 1873.

ASPHALTE (COMPRESSED).

Repairs and Condition on March 1, 1873.

Continually repaired at night. Many holes in the entire surface. Channels much worn. Extensive repairs required this year.
Continually repaired at night. Many small holes in the surface; mostly appeared within the last 6 months. Very extensive repairs required this year.
No repairs done. A few small holes; otherwise in good condition.

No repairs done. Channels somewhat worn.
One or two repairs done. One or two holes, channel somewhat worn. Otherwise in good condition.
One or two repairs made. In good condition.
Trifling repairs. In fair condition.
No repairs. In good condition.
Frequent repairs at south end, where there are many small holes. Many depressions towards north end. Generally in good condition; but not so good as others of this Company.
Trifling repairs. Small holes over the surface. Generally in good condition.
A few small holes. In very good condition.
Trifling repairs. Channels worn; a few small holes. Otherwise in good condition.
Slight repairs. Some short holes. Generally in good condition.
The first pavement laid in the City. Nine very small repairs made. A few holes. Generally in good condition.
Narrow way frequently repaired. Channels deeply worn. Well-marked depression at middle by wheels. Many short holes at east end. Much wear.
One slight repair. A few small holes. Otherwise in good condition.
Slight repairs. Surface worn; otherwise in good condition.
No repairs; in good condition.

TRAVERS (MASTIC).

No repairs. Small depressions on surface; but generally in good condition.

TABLE No. 37.—CITY OF LONDON:—DURATION AND

LIMMER

Locality.	Approximate average width of pavement.	Vehicles in 12 hours in 1872-73. Total.	Vehicles in 12 hours in 1872-73. Per foot of width	Time down at March 1, 1873. Yrs.	Time down at March 1, 1873. Mos.
	Feet.	Vehicles.	Vehicles.	Yrs.	Mos.
Lombard Street	17	2,499	147	1	9½
Moorgate Street	32	6,000	188	1	5¼
Cornhill	23	3,507	—	0	11¾
BARNETT'S					
Moorgate Street	32	6,000	188	1	4½
Carter Lane	8	317	40	0	5¾
Lothbury	43	1,612	38	0	4¾
Bishopsgate St. Within . .	28	6,048	216	0	2½
MONTROTIER					
Princes Street	24	5,628	235	0	6½
SOCIÉTÉ FRANCAISE DES					
Princes Street	21	5,628	268	0	7½

Repair of Asphalte Pavements, 1873—*continued.*

Asphalte (Mastic).

Repairs and Condition on March 1, 1873.

Extensively repaired in last 3 months. Channels worn. Many short holes at middle of way.
Extensively repaired in last 3 months. Considerable indications of surface wear in the middle; a few bad holes and depressions. Otherwise good.
Trifling repairs. Channels slightly worn. Middle of way considerably worn; surface rough.

Asphalte (Mastic).

Extensively repaired. Many small holes, especially at south end, which was laid in very bad weather.
Surface in good condition.
Large portion badly laid, re-laid. Surface variable, some parts smooth, others worn and rough.
No repairs. Numerous holes; and indications of loose structure or of wear. Eastern part rough, indicating inferior material or bad laying. Weather very bad when it was laid.

Asphalte (Compressed).

No repairs. In good condition.

Asphaltes (Compressed).

No repairs. Unusually wavy. Generally in good condition.

In compiling this table, an attempt has been made to gauge the intensity of the traffic by a calculation of the daily number of vehicles by which each street was traversed, per foot of the width of carriage-way. These traffics are placed in the following table, No. 38, for comparison with the loss of thickness, taken at per year, in the last column:—

TABLE No. 38.—CITY OF LONDON:—WEAR OF ASPHALTE PAVEMENTS, 1873.

VAL DE TRAVERS ASPHALTE (COMPRESSED).

LOCALITY.	Vehicles in 12 hours per foot of width.	Time down at March 1, 1873.	Loss of nominal thickness.	
			Total.	Per year.
	Vehicles.	Yrs. Mos.	Inch.	Inch.
Poultry: next kerb .	363	2 2¾	·44 to ·94	·20 to ·42
Do. general surface .	363	"	·375 to ·75	·17 to ·34
Cheapside: next kerb .	298	2 2¾	·81 to ·94	·36 to ·42
Do. general surface .	298	"	⅜	·17
Old and New Broad Streets . . .	103	1 11¼	·31	·16
Queen Street . .	153	1 10	¾	·41
Moorgate Street . .	188	1 6½	·118	·08
Threadneedle Street, east end and centre.	142	1 2¼	1·00 (next kerb)	·87
		1 1¼	1/16 (general surface	·06
Average for general surface . . .	208	—	—	·19
LIMMER ASPHALTE (MASTIC).				
Lombard Street . .	147	1 9½	⅛	·07
BARNETT'S ASPHALTE (MASTIC).				
Moorgate Street . .	188	1 4½	1/16	·05
MONTROTIER ASPHALTE (COMPRESSED).				
Princes Street . .	235	0 6½	½ to ¾	·35 to 1·40

The data contained in this table, No. 38, are too scanty for the purpose of making a useful comparison of the wears of the different asphaltes; but a general relation may be observed between the traffics per foot of width and the vertical wears per year of the pavements in Val de Travers Compressed Asphalte; and the averages are as follows:—

Val de Travers Compressed Asphalte Pavement for Carriage-ways.
Vehicles in 12 hours per foot of width . . . . 208 vehicles.
Loss of thickness per year (say ⅛ inch) . . . . ·19 inch.

It has been argued that the reduction of thickness is the result of simple compression under the traffic, and not sensibly of wear:—an argument which is countenanced by Colonel Haywood, who found that asphalte is not so noiseless after it has been down for two or three months, as it is when first laid, and ascribed the difference to compression and solidification under traffic. Be that as it may, deterioration of the surface is augmented by the action of the traffic; and, ultimately, the abrasion, and the consequent loss of thickness, must proceed in an increasing ratio, for it can scarcely be questioned that the material next the upper surface is originally more durable than what is under it.

With the experience of nearly a year later than the date of his report from which the foregoing data are derived, but without sufficient experience to determine the actual durability of asphalte in the City, Colonel Haywood, after full consideration of all the circumstances, thought "that, without much repair, none of the asphaltes would last more than from four to six years; and that, in the course of from six to ten years, the entire surface of all will have been renewed." In explanation, it is to be said that the repair, however slight, of an asphalte carriage-way pavement, requires that entirely new material should be applied in replacement of that which is in a state of disrepair. That is to say, there is no such operation as the relaying of an asphalte pavement. But, the asphalte that is removed, is utilised by being converted into a liquid asphalte for the formation of footpaths.

In the Appendix, a table, extracted from a recent report of Colonel Haywood, gives the condition of the asphalte carriage-way pavements in the City of London, on the 1st of February, 1877. It may be noted that the works of two of the Asphalte Companies,—the Montrotier, and Barnett's,—no longer exist.

The asphalte pavements in the City of London are all maintained for periods of 16 or 17 years, by contract with the respective companies by whom they were laid.

	Concrete. inches.	Asphalte. inches.	Completed.
VAL DE TRAVERS ASPHALTE, (Compressed.)			
Cheapside	9	$2\frac{1}{4}$	Dec. 1870
Poultry	9	$2\frac{1}{4}$	Dec. 1870
Old Broad Street	6	2	March, 1871
Gracechurch Street	9	$2\frac{1}{8}$	July, 1871
Finsbury Pavement	6	2	Aug. 1871
Moorgate Street (N. end)	6	2	Aug. 1871
Queen Street	9	$2\frac{1}{8}$	April, 1871
LIMMER ASPHALTE (Mastic).			
Moorgate Street (central)	9	2	Sept. 1871
Lombard Street	9	2	May, 1871
Cornhill	9	2	March, 1872
Mincing Lane	6	2	Aug. 1873
BARNETT'S ASPHALTE (Mastic).			
Moorgate Street (S. end)	9	$2\frac{1}{4}$	Oct. 1871
VAL DE TRAVERS (Mastic).			
George Yard	6	$1\frac{1}{2}$	April, 1871

The terms of the contracts for these streets are given in Table No. 39, which follows:—

TABLE No. 39.—CITY OF LONDON:—COST OF ASPHALTE PAVEMENTS.

Laid and Maintained by Contract.

Locality.	Period of maintenance.	First cost per square yard.			Contract cost for maintenance per square yard.	Total cost per square yard.	
		Foundation.	Asphalte.	Total.		Total.	Per year.
VAL DE TRAVERS (compressed).	Yrs.	s. d.	s. d.	s. d.	s. d.	s. d.	s. d.
Cheapside . .	17	1 9	16 3	18 0	2 yrs. free. 15 „ at 1s. 6d. = 22 6	40 6	2 4½
Poultry . .	17	1 9	16 3	18 0	„ „ 22 6	40 6	2 4½
Old Broad St. .	17	1 9	14 3	16 0	2 yrs. free. 15 „ at 9d. = 11 3	27 3	1 7¼
Gracechurch St.	17	1 9	15 3	17 0	2 „ free. 15 „ at 1s. = 15 0	32 0	1 10½
Finsbury Pavement . .	17	1 0	14 3	16 0	2 „ free. 15 „ at 9d. 11 3	27 3	1 7¼
Moorgate St. .	17	1 9	14 3	16 0	„ „ 11 3	27 3	1 7¼
Queen St. . .	17	1 9	14 3	16 0	„ „ 11 3	27 3	1 7¼
LIMMER (mastic).							
Moorgate St. .	17	2 8	13 4	16 0	2 yrs. free. 15 „ at 9d. = 11 3	27 3	1 7¼
Lombard St. .	17	2 8	13 4	16 0	„ „ 11 3	27 3	1 7¼
Cornhill . .	17	2 8	12 4	15 0	„ „ 11 3	26 3	1 6½
Mincing Lane .	17	1 9	10 3	12 0	„ „ 11 3	23 3	1 4½
BARNETT'S (mastic).							
Moorgate St. .	18	2 8	10 6	13 2	3 yrs. free. 15 „ at 1s. 4½d. = 20 7	33 9	1 10½
VAL DE TRAVERS (mastic).							
George Yard .	10	1 9	10 3	12 0	10 „ free.	12 0	1 2½

NOTES TO TABLE.—1. Excavation is not included. 2. The pavements to be given up as good as new.

ASPHALTE PAVEMENT IN MANCHESTER.

A specimen of Val do Travers asphalte, 1,000 square yards, was laid in a street in Manchester, and was down two years. During 17 days in December, 1873, and January, 1874, 61 horses fell, or 3·6 horses per day:—comprising 12 carriage horses and cab horses, 39 lurry horses, and 10 cart horses. The surface was very slippery, and considered to be extremely dangerous except in exceptionally dry weather, or wet weather, and not suitable for the humid atmosphere of the Manchester district. The pavement was replaced with granite.

CHAPTER XVIII.

OTHER PAVEMENTS.

Metropolitan Compound Metallic Paving.—This pavement is formed of blocks of a hard cement, having an iron frame bedded on the top enclosing a certain portion of wood. A specimen was laid in Threadneedle Street, in 1853; it did not prove satisfactory, and was replaced by the proprietors by a new and better specimen, in January, 1854. By the end of the year, it became very unsatisfactory, and was extensively repaired. In June, 1855, it became dangerous, and was finally removed.

Cast Iron Paving.—General Krapp's cast-iron pavement, an American invention, consists of cast-iron frames divided into sections, the divisions being sufficiently close together to prevent the admission of horses' hoofs. The runs and divisions are about 1 inch in width on the top, closely grooved to about an inch in depth. The frames are so designed that, to some extent, there would be a connection and mutual support throughout a paving laid with them. They are laid upon the usual substratum, in the same manner as granite paving, the interstices being filled with gravel, stones, or concrete. The weight of the pavement is 250 pounds per square yard; it costs, when complete, about 18s. per yard. A specimen was laid in Leadenhall Street, early in 1855. By the middle of the year, it got very much out of repair, and it was relaid on concrete, the interstices being filled with concrete; but its condition was not very satisfactory. In February, 1857, it was in an extremely bad state of repair; and in April, it was finally removed, after having been down for a little over 2 years.

The Cellular Iron Pavement.—This pavement consists of cast-iron blocks, 14 inches square, having a flat sole with a downward stiffening rib at each edge, and honeycomb reticulations on the upper side, which are open through the casting. The six-side cells so formed, are oblong, there being six cells side by side, one way, and four the other way of the block. The cells overhang the edges a little, and are the means of interlocking the blocks vertically when the blocks are placed in position. Thus, it was designed, the paving would rest solidly and immovably on a common substratum; and any substance like sand or gravel thrown on the pavement, was to be driven through the cells by the weight of the traffic, and would in fact constitute a continuous packing. This pavement was tried in the Poultry in 1863.

Poletti and Dimpfl's Artificial Granite Pavement.—This pavement is made into blocks, $3\frac{1}{2}$ inches wide, 6 inches deep, 8 inches long, composed of ordinary clay, highly compressed and then burnt. Each block weighs about 12 pounds. The bed for the pavement is levelled and made up with ordinary ballast, and the joints between the blocks, $\frac{1}{4}$-inch wide, are filled up with the same material.

Compound Wood-and-Stone Pavement.—Mr. Newlands, in 1855, tried a system of paving granite sets and wood blocks in alternate courses, for the purpose of lessening the noise of granite pavement. A portion of Great Howard Street, Liverpool, was paved in this manner, and, for a short time, it appeared to answer the purpose. But the wood speedily became depressed below the surface of the granite, either by compression or by wear, and the gaps thus made between the stones had the effect of increasing instead of diminishing the noise.

Concrete Pavement.—This pavement, designed by Mr. Joseph Mitchell, is laid on a bed of Portland-cement concrete, 3 inches deep. The bed may be made with broken stones or with gravel. It is allowed to remain 10 days for consolidation. Paving sets, 2½ or 3 inches wide, and 5 inches deep, are *built* on the bed, with cement mortar, and the joints are filled up with cement grout. The best cement only is used, and the pavement becomes, in fact, a work of masonry 8 inches deep. A piece of this pavement was laid on George IV. Bridge, Edinburgh, in 1866, where the traffic was heavy and continuous; and, after 3½ years of work, it was in excellent order, water running freely off the surface, whilst the wear was scarcely perceptible, and the surface was said to be much freer from mud and dust than ordinary pavement. The cost was said not to exceed that of ordinary pavement made with sets 9 inches deep.

CHAPTER XIX.

COMPARISON OF CARRIAGE-WAY PAVEMENTS.

Comparative Costs.—The experience of wood-pavements of the best construction, and also that of asphalte, for carriage-ways, in England, is of recent origin, and the terms of a comparative statement of their costs must be approximative. The results of various estimates already given, are here grouped together for easy reference, not as matured estimates, nor even as data for direct comparison:—

Granite Pavement, in the City of London, excluding Foundation.

	Duration.	Total cost per year.
Gracechurch Street (page 185)	25 years	7·92d. per square yard
London Bridge (page 184) .	9 ,,	1s. 10¼d. ,, ,,
Five Thoroughfares (page 188)	15½ ,,	1s. 7d. ,, ,,

Wood Pavements, in the City of London, including Ballast Foundations.

| Carey's—six thoroughfares . | 10¼ years | 2s. 6d. per square yard |
| Improved Wood—four do. . | 16 ,, (cont.) | 2s. 4d. ,, ,, |

Asphalte Pavements, in the City of London, including Foundations of Concrete.

Val de Travers (Compressed) . }	17 yrs. (cont.)	1s. 7¼d. to 2s. 4¼d. per square yard
Limmer (Mastic) .	17 ,, ,,	1s. 4¼d. to 1s. 7¼d. ,, ,,
Barnett's (,,) .	18 ,, ,, .	1s. 10½d. ,, ,,

An allowance, at the rate of 1s. 9d. per square yard for the first cost of a concrete foundation, may be made, in estimating the addition to be made to the above-noted costs of granite pavements, to bring them into direct comparison with the other pavements. But, the costs as given for wood and asphalte are only nominal. There is reason for believing that the total cost for construction and maintenance of wood-pavement of the best recent construction will be found to be much less than the costs above given for it; and that the cost for asphalte pavement will be much greater than the costs above given for it.

The cost of Yorkshire paving, 3 inches thick, laid in the City of London, amounted, in 1850, to from 6s. to 6s. 6d. per square yard; and the average cost for repair amounted to about one penny per square yard per year. After remaining down for about 18 years in principal thoroughfares, the paving was removed and relaid in courts or in streets of inferior traffic, where it was estimated it would last from 15 to 20 years longer. The total life was taken at 36 years, which was about the same length of time as the life of granite carriage-way pavement at the same epoch.

But it has recently, in 1875, been estimated that the Yorkshire stone foot-pavements of London, now last only 7 years in the busiest localities, or 12 years in localities of small traffic.* There appears to be an error in the estimate.

The cost of supplying and laying 3-inch Yorkshire flag-stones in Liverpool, before the introduction of Caithness flags, was from 3s. 6d. to 3s. 9d. per square yard.

The following table is given by Mr. G. J. Crosbie Dawson, for the average costs of foot pavements:—

	Per square yard	
	s.	d.
Yorkshire Flags	9	9
Asphalte, Val de Travers	6	0
Blue Brick Paving	4	6
Gravel Paths	2	3

Comparative Slipperiness.—Colonel Haywood, in 1873,

---

* *The Builder*, July 31, 1875, page 679.

made exhaustive observations on the accidents to horses on carriage-way pavements in the City of London:—granite, wood, and asphalte. The wood pavements consisted of the "Improved Wood," and the "Ligno-Mineral" pavements. The asphalte pavement was the Val de Travers (compressed).

The average of 50 days' observations showed that granite was found to be the most slippery, asphalte next, and wood the least. That,—

    Granite was most slippery when dry; safest when wet.
    Asphalte   ,,   ,, damp; ,, ,, dry.
    Wood   ,,   ,, damp; ,, ,, dry.

The numerical data upon which these conclusions are founded, are summarised in the Table No. 40.

TABLE No. 40.—CITY OF LONDON:—COMPARATIVE SLIPPERINESS OF CARRIAGE-WAY PAVEMENTS.

Pavement.	Distance travelled before a horse fell.			
	Dry.	Damp.	Wet.	Average of 50 days.
	Miles.	Miles.	Miles.	Miles.
Granite	78	168	537	132
Asphalte	223	125	192	191
Wood (two kinds) . .	646	193	432	330
Improved Wood . . .	—	—	—	446
Ligno-Mineral . . .	—	—	—	58

Here it is to be observed, that the worst condition of pavement was dry granite, and the best condition was dry wood. The Ligno-mineral wood pavement is obviously far inferior, in point of safety, to the Improved Wood. Improved Wood may be taken as representative of other wood-pavements recently laid, in respect of safety to horses.

Comparative Convenience.—Wood pavements make the least noise; and asphalte pavements make less than granite. Of the wood pavements, the quietest are those which are laid on concrete; and the closer the joints, the quieter and steadier is the motion over the pavement. For, in all wood pavements, the fibre is turned over at the edges into the interspaces; and a series of hollows is established between the blocks, which, in the more widely set pavements, give rise to unpleasant vibration in vehicles passing over them,—like that produced by a stiff spring insufficiently loaded. Henson's pavement is the only one in which such vibration is prevented; and it is no doubt the dryest, the smoothest, and the most silent of wood pavements, whilst it is likely to be amongst the most durable.

Wood absorbs moisture, and is frequently damp when asphalte is dry; but if it be reasonably clean, the dampness does not affect the safety or the comfort of the traffic. Colonel Haywood says that, although some streets in the City have been paved with wood for thirty years, no complaints of offensive smell, or of unhealthiness, have been made to the Commission.

The Committee of the Society of Arts, on *Traction on Roads*, express, in their report,* very strong sanitary objections to wood as a material for pavements. Since that report was written the practice of wood paving has been much improved. As Colonel Haywood puts it, "in confined places, and under some conditions, wood might be objectionable. I have seen it decaying in confined places without traffic."

Printed in the *Journal of the Society of Arts*, June 25, 1875.

CHAPTER XX.

CLEANSING OF PAVEMENTS.

THE dust and mud of the streets of great towns consist generally of four things :—comminuted stone, horse-dung, shoe-iron, and shoe-leather. House-refuse is not an unknown element; it counts for something in the mass of articles that is collected in and removed from the streets and roads. Dr. Letheby, in 1867, analysed dry mud from the streets of the City of London,—dried by exposure for many hours to a temperature of from 266° to 300° Fahr. At the same time, he analysed, for comparison, well-dried fresh horse-dung, and common farm-yard dung. The results of the analyses, for the mud of stone-pavements, are given in table No. 41.

TABLE No. 41.—COMPOSITION OF MUD FROM STONE-PAVED STREETS, HORSE-DUNG, OR FARMYARD DUNG.

Dried at 300° Fahr.

Constituents.	Fresh horse-dung.	Farm-yard dung.	Mud from stone-paved streets.		
			Maximum organic (d. y weather).	Minimum organic (wet weather).	Average.
Organic	Per cent. 82·7	Per cent. 69·9	Per cent. 58·2	Per cent. 20·5	Per cent. 47·2
Mineral	17·3	30·1	41·8	79·5	52·8
	100·0	100·0	100·0	100·0	100·0

CLEANSING OF PAVEMENTS.

The higher proportion of mineral matter in wet weather proves that in such weather the abrasion of stone and iron is greatest. Dr. Letheby estimated that the average proportions of stone, iron, and dung in the muds were:—

Horse-dung	57 per cent.
Abraded stone	30 ,,
Abraded iron	15 ,,
	100 ,,

The mud was so finely comminuted that it floated freely away in a stream of water.

In the mud of wood pavements, the proportion of organic matter in the dried mud was larger than in the mud of stone pavements. It amounted to about 60 per cent.

The amount of moisture in the street mud varied according to the state of the weather:—

STONE PAVEMENT.   Moisture.

In the driest weather	rarely less than 35 per cent.
In ordinary weather	,, 48½ ,,
In wet weather	,, 70 to 90 ,,

By far the greatest proportion of the detritus of macadamised roads consists of the worn material of the road; and the important principle was early revealed by experience, "that the oftener that streets are cleansed, the less is the mud which is created and removed, whilst the attendant expenses are by no means increased, and the roads are kept in a better state of preservation." This principle is, besides, logically deducible from the fact that the worn particles, if left on the surface, act as a grinding powder under the wheels and the horses' feet, to reduce to similar powder the surface of the road; and that the mud which is formed with the detritus, when rain falls upon and is mixed with it, operat like a sponge in retaining the moisture upon the surface. The upper crust as well as the substratum, under these circumstances, become saturated with moisture, softened, and "rotten;"—just as a gravel footpath, hard and solid in ordinary weather, becomes sodden and pulpy when it is lapped for a time by a covering of half-melted snow. The macadamised carriage-way, thus reduced, is exposed to rapid deterioration by the traffic, which increases in a highly accelerated ratio with the period during which the road is left uncleansed. The statistics of cleansing unanimously support these conclusions. Shortly after Sir Joseph Whitworth introduced his street-scraping machine in Manchester, it was ascertained from calculations made by the municipal authorities on the relative advantages of machinery and manual labour, that, by cleansing the macadamised streets with the Whitworth machine, three times a week, the quantity of mud produced on the surface was only one-fifth of what was produced when they were swept by hand twice in three weeks, and only one-thirteenth of what was produced when they were swept once a week. The following are the statistics referred to:—

MANCHESTER:—Area of district, 5,500,000 square yards.

	Loads removed.	Average area swept to produce a load.
		Square yards.
Swept by machine three times a week	1,285	4,388
Swept by manual labour twice in three weeks. Township, 1841-42	6,400	859
Swept by manual labour once a week. Township, 1838-39	17,000	343

There is a want of harmony in this statement, from which it appears that by hand-scraping once a week more mud was collected than by scraping once in 10 days.

There was probably a difference of circumstances, as there was an interval of three years between the two series of observations. Be that as it may, it is obvious that the more frequent cleansing performed by the machine, was instrumental, not only in keeping the streets in better condition, but in reducing the tear and wear of the streets, as represented by the decreased quantity of detritus which was generated and removed.*

When a road is very soft, it should not be touched; or, at the most, only the loose liquid mud should be removed, by light sweeping, with, at the same time, a copious use of water. When a road is in a soft condition, and is covered with a thick tenacious mud which adheres to the stones, it is a common opinion that the use of water injures the road by softening the surface, and causing a removal of the material. But the truth is exactly the reverse. A road in this condition cannot be effectually cleansed without water:—by the free use of water, the mud is softened, is liquefied, and is easily removed without impairing the integrity of the surface, or removing the useful grit.

The difficulty and cost of cleansing streets effectually by hand-labour, are materially diminished by the substitution of the sweeping machine. The Whitworth machine has been much employed for cleansing streets. The apparatus consists of a series of broad brooms, usually about 30 inches wide, attached to a pair of endless chains which turn upon an upper and a lower set of pulleys suspended in a light wrought-iron frame, behind a cart, the body of which is near the ground. As the cart-wheels revolve, they communicate, by gearing, a rotary motion to the pulleys carrying the endless

---

* See a paper "On the Present State of the Streets of the Metropolis," *Proceedings of the Institution of Civil Engineers*, 1843; vol. ii. page 202.

chains and the brooms attached to them; the brooms being sufficiently lowered so as to bear upon the surface of the street, they successively sweep the surface, and carry up the soil on an inclined carrier-plate, over the top of which it is dropped into the receptacle. The oblong arrangement of brooms, effected by the employment of the endless chains, contrasts favourably with a simply circular or revolving broom; for the ends of the brooms move most rapidly in the act of revolving on the lower pulley, whilst they are in contact with the ground, and they move towards the upper end at a less speed, whilst conveying the soil up the incline. In turning over the upper pulley, the velocity of the ends of the brooms is again raised to the maximum, and the mud is at this place stripped from them by a bar called a "doctor." If, for instance, the horse travels at the rate of 3 miles per hour, the speed of the brooms, whilst sweeping the ground, is 9 miles per hour; whilst passing under the doctor, 6 miles per hour; whilst they convey the mud up the incline at 3 miles per hour. The brooms are raised and lowered by hand, by means of a coiled spring and chain. The spring relieves the pressure of the brooms upon the ground, according to the inequalities. A strong horse is required to work and to draw the machine.

The area of surface swept daily in Manchester, in 1847, by one of Whitworth's machines, with one man and one horse, varied from 14,000 to 20,000 square yards. The quantity removed amounted to from one to three loads per acre, according as the surface was dry or wet.* The cost of working each machine was from 10s. to 15s. per day, exclusive of all the expenses attending the disposal of the sweepings, &c. Sweeping continuously, each machine

* "Mechanical Street Cleansing," *Proceedings of the Institution of Civil Engineers*, 1847; vol. vi. p. 431.

would cleanse, in a day, a strip of roadway 30 inches wide, and 20 miles long, making an area of 29,334 square yards. As a fact, more than 30,000 square yards have been swept in a day. But, a third, at least, of the whole time that the machine is at work, is consumed in conveying it to and from the place of deposit, and unloading. During the second half of the year 1843, each machine swept an average of 18,189 square yards per day, in Manchester. Against this performance it appears, from official returns, that in the years 1840—42, each able-bodied man employed swept, on an average, 1,190 square yards per day, and each pauper swept about 600 square yards. From these data, it appears that the effective work of one machine equalled the work of 17 able-bodied sweepers. The comparative costs may be placed as follows:—

Sweeping by hand:—
   Seventeen sweepers at 2s. per day .   34s.
   Cart, horse, and driver . . .   8s.   42s. per day.
Sweeping by machine, say . . .   —   13s. „

      Difference, saving by machine . . 29s. „

showing that the cost by machine was less than one-third the cost of manual labour.

In Salford, the results of comparative trials of sweeping by hand and by machine, showed that the cost of cleansing was—

		Square yards per day.	Per square yard.
By hand . .	£674 8s. 8d. sweeping	15,904 .	10·18d.
By one machine .	£195 9s. 5d. „	14,165 .	3·31

In Birmingham, according to Mr. Pigott Smith, the extent of surface of urban carriage-way, swept by eight machines, was 254,000 square yards. The whole of this area was macadamised, and was kept well cleansed by the eight machines. The area of surface averaged about 32,000 square yards for one machine, which would have been

CLEANSING OF PAVEMENTS. 275

sweepings were washed, to separate the refuse from the strong matter mixed with it. It was found that one-third of that which was taken up dry, consisted of coarse grit, which would have been useful on the road; whilst one-twelfth part only of that which was removed in the form of liquid mud was stony matter, and it was so completely pulverised, as to be of scarcely any use. After the two portions of the road had been cleansed, the difference between them was very striking: that which was swept whilst dry was still covered with adhesive mud, which, together with the stones to which it adhered, was lifted by the wheels, the whole road being rough and uneven, whilst the portion which had been swept after watering was smooth and even. On the 24th, both portions were again swept, but only one quarter as much dirt was taken from that which had been water-swept as from that which, being in a sticky condition, had been swept without the addition of water. On the 26th, rain fell, and three times as much slop was taken off the part of the road which had not been swept on the 22nd. The preservative effect of water-sweeping by mechanical means, was demonstrated by the decidedly better condition of that portion of the road so cleansed."

The detritus of the material of a granite pavement constitutes but a very small proportion of the total quantity of mud-forming dust. Colonel Haywood exemplified this proportion in an interesting manner, taking the instance of the granite pavement of London Bridge,—3-inch Aberdeen granite sets,—which was removed in 1851, after having been down nine years. The average loss of granite over an area of 3,950 square yards, he estimated, was equal to 2 inches of vertical wear. The total volume of granite worn away was, therefore, about $219\frac{1}{2}$ cubic yards, assuming that the surface was a continuous mass of granite, though there was of course a considerable superficial area of joints. Assuming that the granite worn off was reduced to the state of fine powder, it was increased in bulk probably one half, and its volume had been (219½ × 1½ =) 329¼ cubic yards. Adding 5 per cent. for the loss upon stones removed and replaced from time to time, the total quantity worn off and reduced to powder, and carried away, mixed with the dust of the street, and mud, would only have amounted to 345·7 cubic yards for nine years, equivalent to a wear of ·105 cubic yard—about a tenth of a cubic yard—per day. Whereas, the quantity of dust removed, daily, in dry, calm weather, was from 3 to 3½ cubic yards,—over thirty times as much as the granite detritus. So much for horse-droppings and shoe-leather, which must have constituted 29-30ths of the total accumulation,—independent of the contributions of house-refuse, in the inhabited streets.

Seeing that, according to the results of these observations, the granitic element of the refuse constituted little more than 3 per cent. of the whole, the evidence of the exceedingly greater wear of a macadam surface is strikingly exhibited in the contrast of the comparative costs for cleansing and watering the two kinds of surface. Mr. Royle states that, in the spring of 1870, there were laid in Chester Street, Manchester, 10,000 square yards of granite pavement, and an area of equal extent of macadam. The condition of each surface was maintained uniform throughout the period of trial, by sweeping and watering, for which the respective costs were 9¼d. per square yard, and 5s. 2½d. per square yard,—in the ratio of 1 to 7. The average costs throughout the year, ranged in the ratio of 1 to 5. The great bulk of the excess for macadam, compared with pavement, must have been due to the vastly augmented wear of granite, since the dust and mud from other sources must have been the same—as a constant quantity—for both pavements.

CLEANSING OF PAVEMENTS.

In Reading,* which has about 23 miles of macadam roads, 17 miles are watered twice a day; for which purpose 11 watering-carts and barrels, and 4 hand-watering machines, have been employed. One cart waters a length of 5,962 lineal yards, having an area of 23,849 square yards, twice a day, at a cost of 8s. for horse, cart, and man, and 1s. 5d. for maintenance of cart, and harness, and shoeing; total, 9s. 5d. per day. With one hand-machine, 23,740 square yards are watered twice daily, at a cost of 2s. 10d. each for two men, and 7d. for maintenance; making a total daily cost of 6s. 3d.

	Area watered twice daily.	Cost per day.	Water delivered per square yd.
One water-cart	23,849 square yards	9s. 5d.	0·51 gallons
One hand-machine	23,740 ,,	6s. 3d.	1·30 ,,

Watering the Streets of the City of London with Jet and Hose.—In 1867, and 1872, experiments were made in washing the carriage-ways of Cheapside and the Poultry when laid with granite pavement, and afterwards with asphalte pavement. The asphalte pavement of Old and New Broad Streets was likewise washed experimentally.

Hydrants were erected in Cheapside, at distances of 133 feet apart, and in Broad Streets at 140 feet apart. Six men were employed at each washing of Cheapside in the first trial, and 10 men in the second trial; for which two jets of water were used during the time of the first trial. The work of the 10 men in the second trial was thus distributed:—four in playing the jets, two in moving the hose from place to place, and four with brooms in sweeping the surface of the asphalte, and keeping the channels free from straw and larger refuse. These men also used "squeegees" (india-rubber sweeps) to dry the surface of the asphalte. Of the six

* Mr. Ellice-Clark's *Report*, pages 25, 26.

men employed on the granite pavement, four played the jets, and two moved the hose, and plied the brooms.

TABLE No. 42.—WASHING CARRIAGE-WAY PAVEMENTS WITH JET AND HOSE.

Results of Experiments in the City of London.

Particulars.	1867. Granite pavement. Cheapside, &c.	1872. Asphalte pavement.	
		Cheapside, &c.	Broad Streets.
Length . . . . feet	2,000	2,000	1,436
Area of pavement washed square yards	9,792	10,353	3,671
Time washing . . .	2 h. 19 m.	2 h. 4 m.	1 h. 3 m.
Area washed per hour square yards	4,220	5,000	3,500
Water consumed . gallons	19,500	19,726	9,786
Do. do. per square yard . . . gallons	1·99	1·90	2·66
Cost for labour . . .	9s. 10¾d.	9s.	4s. 10¾d.
Cost for water, at 6d. per 1,000 gallons	9s. 9d.	9s. 10¼d.	4s. 9d.
Cost for labour and water, one day's washing . . .	19s. 7¾d.	18s. 10¼d.	9s. 7¾d.
Cost for labour and water, per square yard; . pence	·0241	·0218	·0315
Cost for labour and water, per square yard, per year; pence	7·510	6·823	9·859
Add for supervision, preparing apparatus, wear and tear, &c., 20 per cent. . . .	1·500	1·377	1·972
Total cost . . .	9·010	8·200	11·831

The granite pavement was more thoroughly cleansed by hose than in the ordinary way, by scavengers. The cost of washing asphalte was only 3 or 4 per cent. less than that for granite.

Mr. T. Lovick's Experiments in Washing the Streets of London.—In evidence before the General Board of Health, in 1850, Mr. Lovick stated that in surface-cleansing by a jet and hose, illustrated in Figs. 57 and 58, the quantity of water consumed was at the rate of rather less than 1 gallon per square yard of surface of carriage-way, working at "extremely low pressures." He mentions experiments made by Mr. Lee, at Sheffield, with "very high pressures," in which Mr. Lee effected the cleansing with a consumption of less than a third of a gallon per square yard. Mr. Lovick estimated that the foot-paths could be cleansed with a consumption of half a

Fig. 57.—Cleansing the Surfaces of Streets.

gallon of water per square yard. He adds that the daily cleansing could be effected at a cost of 3d. per house per week; exclusive of the cost for water; for which he allowed 28 square yards of carriage-way and 16 square yards of footway, per house. At this rate, the first item of cost per square yard of surface would amount to 3½d. per square yard per year. For water, the cost at 6d. per 1,000 gallons, would be 1½d. per square yard per year. The total cost would be 5*d*. per square yard per year.

Mr. Lovick, at the same time, estimated that the cost of cleansing by ordinary scavenging was nearly double

Fig. 53.—Cleansing close Courts and Alleys.

the cost by jet and hose, whilst the jet had been shown to be much more efficacious in removing evaporating matter and filth.*

* He was able "not only to cleanse the pavements by this means, but also to cleanse the walls from urine stains and other filth."

It has been found, as the result of special observations, that,—

>For 4 loads of mud or slop removed from a macadam surface,
>1 load is removed from a granite pavement; and
>$\frac{1}{4}$ load is removed from a wood pavement.

Cleansing in Paris.—In Paris, since 1855, many attempts have been made to substitute sweeping machines for hand labour. But the inconvenience attending their use—the need for a depôt to hold them when not in use, the need for horses to work them, and the difficulties of continuously employing them in frequented thoroughfares,—have hitherto prevented their being generally employed. One of these machines, constructed by M. Tailfer, which worked very well, performed the work of 8 or 10 sweepers. It is fitted with a revolving brush, 5 feet 8 inches long, placed obliquely, and sweeping a breadth of 5 feet at one passage.

In some comparative trials which were made, one man, with a good brush, swept 720 square yards of macadam per hour, dry, lightly watered, without heaping the mud; 480 square yards covered with liquid mud; 360 square yards covered with thick mud. On a paved street, he swept from 600 to 820 square yards per hour.

M. Tailfer's machine performed the work of 7 men in the first case just recorded, 10 men in the second case, and 12 men in the third case; but, in the third case, the work was not perfect. Taking all charges into the account, it appeared that the work done by the machine cost about the same as the work done by hand, but it was done much more expeditiously.

CHAPTER XXI.

MOUNTAIN-ROADS.

The first thing to be observed by the engineer in laying out a mountain-road, is that he should personally examine every possible route before deciding upon the line to be followed. An old-established line of traffic should not be hastily abandoned for one that presents greater engineering facilities, and apparently equal commercial advantages. Major James Browne[*] mentions an instance where, in one district of the Punjab, a new road was made from the hills to the plains, which, it was thought, would at once draw away all traffic from the old, circuitous, and greatly more difficult line. Nevertheless, the old road is used, and the new one is not used, for the simple reason that, whereas both lines are thickly wooded, the trees on the new line are acacias and thorn bushes, whilst those on the old line provide excellent forage for the beasts of burden, for five days' march. Major Browne adds, that the natives of India have generally had good economical and local reasons for the selection of their old lines of traffic.

The broad principle most generally applicable in the selection of mountain-roads, is, according to Major Browne, —" The best line for a mountain-road is that on which the total sum of the ascents and of the descents between

---

[*] "On the Tracing and Construction of Roads in Mountainous Tropical Countries."—*Proceedings of the Institution of Civil Engineers*, 1873-4; vol. xxxviii. page 67.

the extreme points, is the least." At elevations exceeding 8,000 feet, attention is to be given to the action of snow in winter; for want of such consideration, roads, otherwise well laid out, are rendered impassable, for many weeks, by avalanches. In the Himalayas, the northern slopes are thickly wooded, where the southern slopes are bare. The wooded side should be selected; for the trees break the force of the rain as it falls, and the road is better protected than it would be on the bare hillside. Where the road-trace crosses a deep chasm formed by a river, the approaches should be laid down stream, to have the benefit of the slope of the

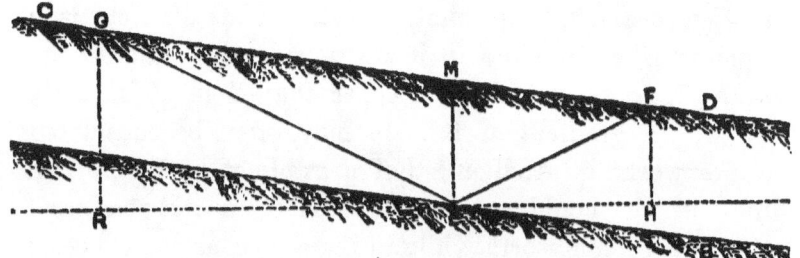

Fig. 59.—Road-Trace approaching a deep Chasm.

river bed. Major Browne illustrates this proposition,— by drawing the horizontal line, R H, Fig. 59, through the point E, where the road is to cross the sloping bed of the river, A B, parallel to the bank, C D, of the vertical height E M. Draw E F and E G through E, making equal angles with the horizontal line. The down-stream trace clears the bank in the distance E F, which is shorter than the up-stream trace E G. The distinction, though obvious, is frequently overlooked.

Heavy expenses and difficult drainage are often avoided by the use of zig-zags in the route. In marking out the formation-level, cuttings which would exceed 10 or 15 feet in depth should be avoided, if possible. In the stony soil of a hilly country, no trustworthy information is to be had by boring, for rock of the toughest and hardest description may crop up where least expected. In tracing a Himalayan cart-road, length is the main thing sought for, to surmount the immense heights met with. Every possible foot of rise should be gained, and never lost.

The steepest admissible gradient for an unmetalled mountain cart-road in India, is usually considered to be 1 in 18, or 5·55 in 100. Where the trace rises steadily, the gradient is broken at every 500 or 600 yards by 100 feet or so, of slight counter-slope, not merely to ease the cattle, but also to break the drainage.

Mr. Dobson, writing from his experience of road-making in New Zealand,* says that, whilst, in level districts, it is sometimes worth while to incur considerable cost for the sake of avoiding inclines steeper than 1 in 30, in hilly country, a gradient of 1 in 10 may often be considered a fair working gradient; that a gradient of 1 in 9 is a limit which should not be exceeded for a cart-road, and that it is seldom worth while to increase constructive costs for the sake of attaining gradients flatter than 1 in 22.

In India, the usual width of a mountain cart-road varies from 18 feet in open ground, to 12 feet along cliffs or in very difficult places; the maximum gradients vary from 1 in 18 to 1 in 25. The metalled road-surface in cutting along a mountain-side, is usually sloped inwards from the outside, at the rate of 1 in 18. The metalling consists of a 9-inch layer of broken granite, kunkur rock (a concretion of carbonate of lime), or coarse slate shingle.

The cost of the construction of a road in India, exclusive of metalling, according to Major Browne, may vary from £1,500, where the bridging is light and labour is abundant, to £2,700 per mile, where the excavation is heavy and the labour is scarce. An addition of £300 per

* "Pioneer Engineering," page 61.

mile is to be made for metalling, which would be at the rate of about 7*d.* per square yard. A portion of the Hindustan and Thibet road,—an unmetalled mule-road,— 7 feet in width, with gradients of from 1 in 6 to 1 in 4, cost about £500 per mile.

The Grand Trunk road, in India, 837 miles in length, constructed to connect distant parts of the Bengal Presidency, has been in use for upward of 25 years. It is raised at every part, about 18 inches above the greatest known height of inundations. The standard width of the surface was, in the first instance, 30 feet; this was afterwards increased to 40 feet, with side slopes of 4 to 1. The central portion, 16 feet wide, is metalled with broken granite, or kunkur, which is laid 8 inches thick, and is rolled or beaten down to 6 inches, by which means a smooth road is formed. When reduced by wear to a thickness of 4 inches, which is generally done in three years, the surface is picked over to the depth of half an inch, and a thickness of 4 or 5 inches of new metal is laid upon it; and it is again rolled or rammed, and at the same time well watered, to bind and consolidate the new material. It is calculated that the whole depth of metalling is renewed every six years. The cost of the road is about £500 per mile, exclusive of large bridges. The cost of laying the metalling is stated as £162 2*s.* per mile, or 6·22*d.* per square yard; the cost for repair and maintenance of metalling is £23 12*s.* per mile, or 1·38*d.* per square yard per year.

Fig. 60.—Roads in India:—First-class Road in Embankment.

The usual section for a first-class road in India, like what has just been described, is shown in embankment,

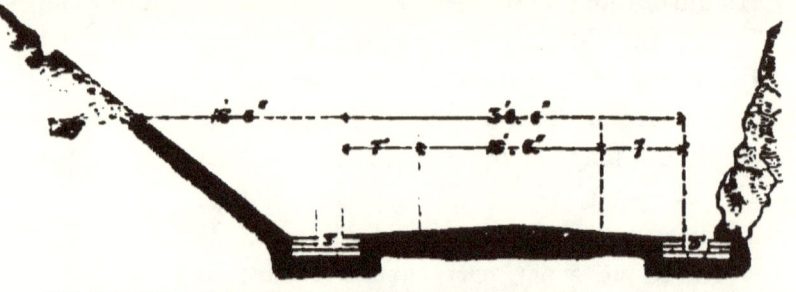

Fig. 60.—Roads in India :—First-class Road in Heavy Cutting.

in Fig. 60; and in heavy cutting in Fig. 61. Hill roads are shown in section in Figs. 62, 63, and 64.

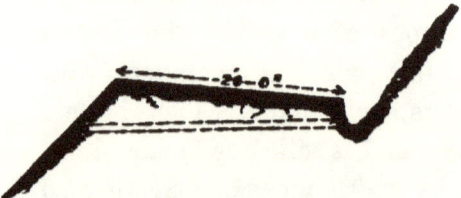

Fig. 62.—Roads in India :—Hill Road.

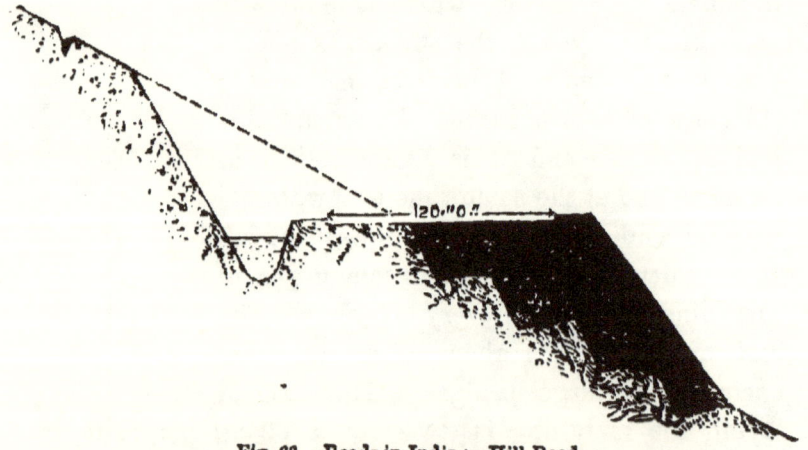

Fig. 63.—Roads in India :—Hill Road.

The vertical lines of cliffs present the most formidable obstacles to the formation of a road along their faces. In

Fig. 65, if A B C be the section of the cliff, and the rock be sufficiently hard and stiff, a half-tunnel like D E F may be

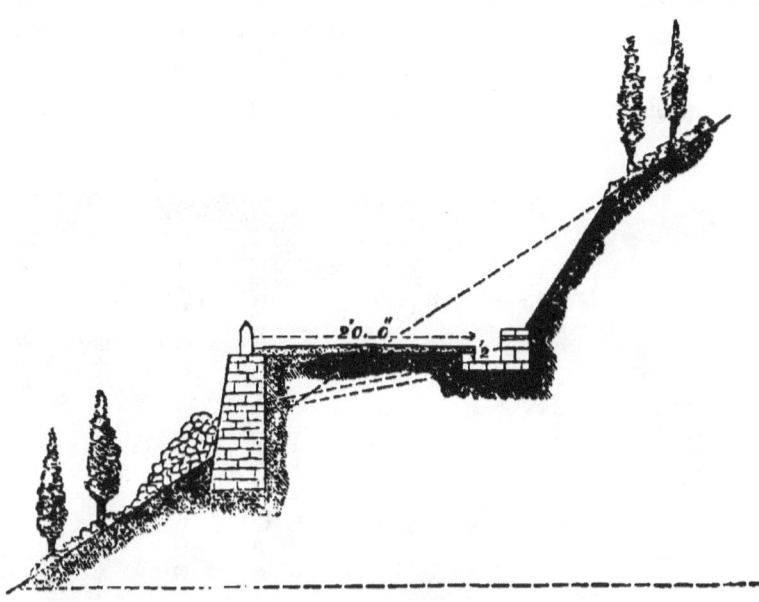

Fig. 64.—Roads in India:—Hill Road.

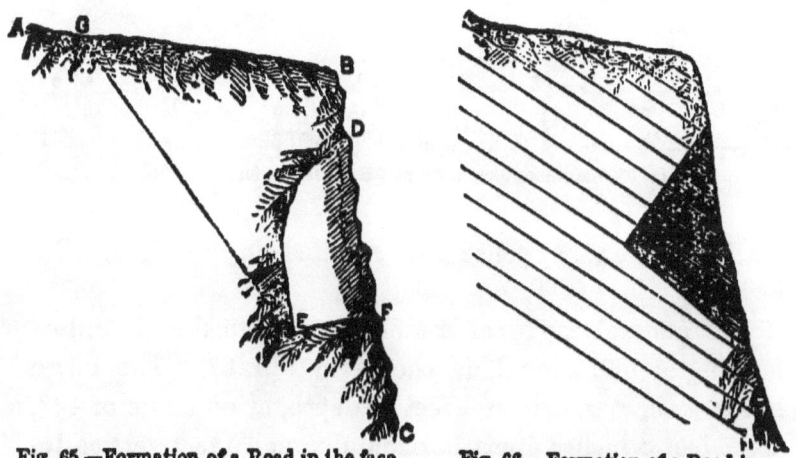

Fig. 65.—Formation of a Road in the face of a Cliff.

Fig. 66.—Formation of a Road in the face of a Cliff.

blasted out; but if it be too soft and rotten to admit of this being done, the best plan, if the cliff be of any great height, D F, above the formation level, is to blow out the whole piece G E F by a large mine at E. Mining should not, as a rule, be employed where there is a chance of the strata being blown out downwards, according to the dip; for a piece may be blown out, like the shaded portion, Fig. 66, when

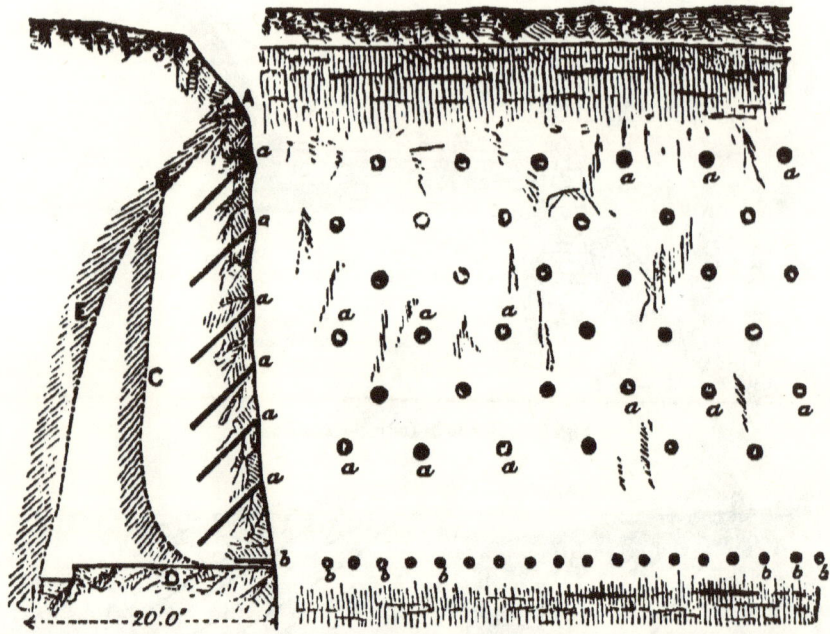

Fig. 67.—Mode of formation of a Road in the face of a Cliff.

much time and expense are entailed in rectifying the level.

The general mode of attacking a vertical cliff, and of forming a half-tunnel, is shown in Fig. 67. The large blasts, *a, a, a, a*, driven 8 feet in depth, at an angle of 45°, are 7 feet 3 inches apart horizontally, and 5 feet vertically. The small holes *b, b*, &c., 3 feet apart and 3 feet deep, which are not fired, serve to determine and facilitate rupture at the proper level. These blasts, when fired, generally blow out or loosen a piece like A B C D. The remaining space, D E F, is blown out in the same manner.*

"It sometimes happens," says Mr. Dobson, "that advantage can be taken of the natural stratification to economise work in a long side-cutting. This was done by the Author in the case of a road over Evan's Pass, at Port Lyttelton, New Zealand. The descent of the Pass was on the side of a long volcanic spur, formed by a succession of lava streams, dipping at an angle of 1 in 12, the lower part of each lava stream being hard volcanic rock, whilst the upper portion was soft and easily worked.

"The line was originally set out with a gradient of 1 in 17, which would have entailed a series of cuttings through the hard rock, and retaining-walls in front of the softer portions. By altering the gradient, however, to that of the lava-streams, a solid floor was obtained throughout, the retaining-walls were dispensed with, and the excavation was made chiefly in soft material. The alteration effected considerable saving in time and first cost, as well as in the cost of maintenance."†

* These particulars are derived from Major Browne's paper, already referred to.

† See Mr. Dobson's work on *Pioneer Engineering* for valuable information on the tracing of roads in mountainous districts.

CHAPTER XXII.

RESISTANCE TO TRACTION ON COMMON ROADS.

In investigations into the resistance to wheels on common roads, it is usual to construct a diagram showing a wheel on a horizontal plane in contact with a stone, over which it is to be pulled; and the force required to pull it over the obstacle is calculated. But this is not the kind of resistance worth investigating, and it certainly has no relation to the kind of resistance which is usually opposed to the wheels of vehicles on inferior roads. The resistance is that of a medium distributed over the submerged portion of the circumference of the wheel, in advance of the perpendicular line drawn from the centre of the wheel to the plane of the road. Let $a\,o\,b$ be such a wheel drawn over the horizontal surface $c\,d\,e$ of the road, in the direction $o f$; and let the road be of such a consistency that the wheel penetrates to the depth $d\,b$ below the surface, leaving a track $b\,g$ behind it. The arc $b\,c$ is the submerged portion of the circumference, and it may be assumed to be identical with the chord of the arc, $b\,c$. Now, the resistance is dis-

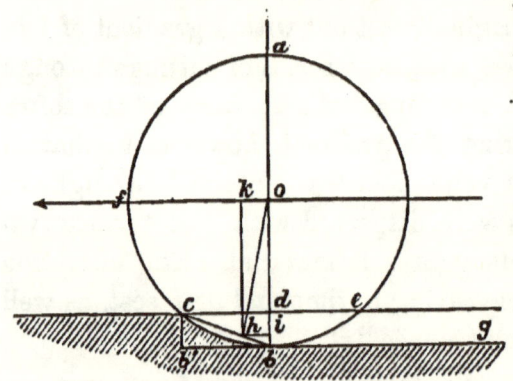

Fig. 68.—Rolling Resistance to Traction.

tributed over the surface bc, and it may be taken as acting on this surface perpendicularly to the plane of the road, or vertically and directly opposed to the gross weight consisting of the weight of the wheel and the load upon it. To simplify the investigation, let it be supposed that the upper portion of the road is homogeneous, as clay or sand; then, the resistance to penetration is nothing at the surface, and it increases as the depth; and the upward resistance along the line of submersion bc, is a maximum at b and it vanishes at c, and the varying intensity of the graduated pressure may be represented by an isosceles triangle, of which the centre of gravity h, situated at one-third of its length, bh, from the base b, is also the centre of resistance, and therefore also the centre of pressure under the load; and the radial line oh is the resultant of the pressure of the load, measured in force and direction by the vertical oi, and the traction force, measured by the horizontal line hi, or ok. But the vertical oi may be taken as equal to the radius ob, and the horizontal hi may be taken as one-third of the semichord of submersion cd; whence the simple proportion,—

Load : tractive force : : ob : cd
:: radius of wheel : $\frac{1}{3}$ semichord ;

and the resistance to traction is equal to the product of the load by the third of the semichord, divided by the radius of the wheel.

But the length of the semichord cd may be more easily determined by calculation from the measured depth of submersion db. It is equal to the square root of the product of the segments into which the diameter ab is divided by the plane of the road cde, or to $\sqrt{ad \times db}$; and the whole of the calculation is embraced by the equation,—

$$\text{Tractive force } ok = \tfrac{1}{3} \times \frac{W\sqrt{ad \times db}}{ob}. \qquad (1.)$$

The equation is, no doubt, applicable, with a sufficient degree of accuracy, for any real needs, for calculating the resistance of gravel, loose stones, soft earth, or clay.

The work done in compressing the material of the road is easily indicated diagrammatically, by supposing the wheel to advance through a space equal to the semi-chord $c\,d$, or the length of submersion. Thus, in Fig. 69, the wheel $a\,b$ is supposed to roll forward and to occupy the position $d'\,b'$. The work done in compressing the road is proportioned to the four-sided area $b\,c\,d'\,b'$, comprised between the circumferential segments $b\,c$ and $b'\,c'$; and this area is, by the properties of the circle, equal to the original rectangular area $b\,d\,c\,b'$.

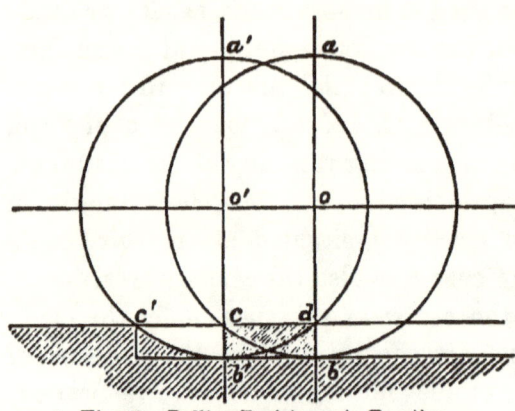

Fig. 69.—Rolling Resistance to Traction.

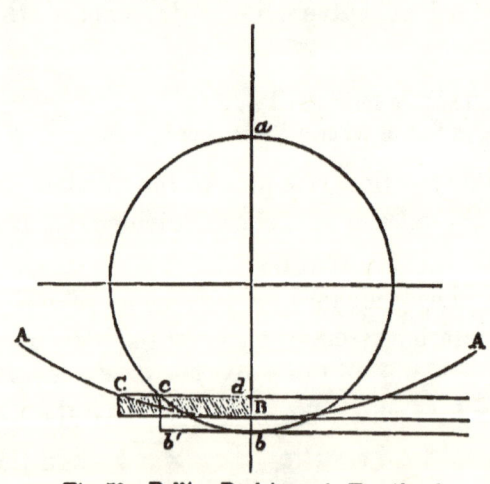

Fig. 70.—Rolling Resistance to Traction.

Now, suppose a wheel ABA, Fig. 70, of larger diameter with the same gross weight, to travel over the same surface. It is obvious that, if it could sink to the same depth, $d\,b$, as that for the smaller wheel, the length of immersion, $d\,c$, would be increased, and the rectangle $db \times dc$, representing work, would be greater than that performed by the smaller wheel in the first example. Such a supposition cannot be admitted: the depth of immersion d B, for the larger wheel, must be less than that, db, for the smaller wheel, though the length of immersion d C, must be greater than that, dc, for the smaller wheel; but not so much greater as if the wheel were sunk to the first depth db.

In fine, larger wheels sink less, but spread more, into the surface, than smaller wheels, in such proportion that the area of the rectangle representing work of submersion is constant for all sizes of wheels. In this instance, accordingly, the rectangle $db \times dc =$ the rectangle d B $\times d$ C.

It might be thought that, on this principle of the constancy of the work of submersion, in a soft road, the resistance to traction must be the same for all diameters of wheels. But, as the rectangle of work is spread over a longer space, d C, for the larger wheel, than the space, dc, for the smaller wheel, it follows, on the contrary, that the resistance or force of traction varies in some proportion inversely as the diameter, being less as the diameter is greater. This conclusion accords with experience; but, though the actual law of variation may not be strictly deducible in the line of reasoning here traced, it is nevertheless useful to carry the reasoning to its logical conclusion. Let a and A be the diameters respectively of the smaller and the larger wheels; b and B the depths of immersion; and c and C the lengths of immersion, or dc and d C respectively. As already stated, the areas of immersion are equal to each other; or,

$$bc = BC. \qquad (2.)$$

Also, the values of c and C are, by the properties of the circle, expressible by the products $\sqrt{ab}$ and $\sqrt{AB}$, for all cases that need occur in practice; and, by substitution in the equation (2),

$$b\sqrt{ab} = B\sqrt{AB}; \qquad (3.)$$

and, squaring both sides,

$$ab^3 = AB^3. \qquad (4.)$$

Finally, extracting the cube root of each side of this equation (4), the equation (5) is obtained,

$$b\sqrt[3]{a} = B\sqrt[3]{A}, \qquad (5.)$$

which may be developed into the proportion,—

$$b : B :: \sqrt[3]{A} : \sqrt[3]{a}; \qquad (6.)$$

showing that the depth of immersion varies as the cube root of the diameter. But, as $bc = Bc$, and $b : B :: c : c$, then,

$$C : c :: \sqrt[3]{A} : \sqrt[3]{a}, \qquad (7.)$$

showing that the length of immersion is as the cube root of the diameter. It has already been seen that the force of traction is as the length of immersion; therefore, finally,—

The circumferential or rolling resistance of wheels to traction on a level road, is inversely proportional to the cube root of the diameter.

On this principle of resistance, it follows that, to reduce the rolling resistance of a wheel to one-half, for instance, the diameter must be enlarged to eight times the primary diameter.

The deduction of M. Morin, mentioned at page 51, that the resistance varies simply in the inverse ratio of the diameter of the wheel,—so that, for example, a wheel of twice the diameter would only incur half the resistance,—has been generally accepted. But this deduction is not supported by the foregoing analysis of forces, and there is good reason for renouncing it, in the more recent experiments of M. Dupuit. He placed model wheels or rollers of various diameters at the summit of an inclined plane, succeeded by a horizontal plane, on which they rolled down by the force of gravity and arrived at a state of rest after having expended the energy acquired in falling through the height of the plane. From these and other experiments he drew the following deductions:—

On macadamised roads in good condition, and on uniform surfaces generally,—

1. The resistance to traction is directly proportional to the pressure.
2. It is independent of the width of the tyre.
3. It is inversely as the square root of the diameter.
4. It is independent of the speed.

M. Dupuit admits that on paved roads, which give rise to constant concussion, the resistance increases with the speed, whilst it is diminished by an enlargement of the tyre up to a certain limit.

M. Debauve* submits the following data for resistance of vehicles on common roads, in which it may be remarked that the resistances decrease in chronological order,— owing, no doubt, as he says, to the progressive improvements in vehicles and in roads. The data are here given in English measures:—

1st.—*On Metalled Roads:*

	Per cent.			Per ton.
Count Rumford	7·7	or 1-13th of the wt.;	or	172·5 lbs.
Gordon	6·2	or 1-16th „ „	„	139 lbs.
Coste and Perdonnet	2·9	or 1-35th „ „	„	65 lbs.
Navier	2·2	or 1-46th „ „	„	49·3 lbs.

2nd.—*On Paved Roads:*

	Per cent.			Per ton.
Rumford	5·5	or 1-18th of the wt.;	or	123·2 lbs.
Navier (walking pace)	4·0	or 1-25th „ „	„	89·6 lbs.
Do. (at a trot)	7·1	or 1-14th „ „	„	159·0 lbs.
Coste and Perdonnet	3·3	or 1-30th „ „	„	73·9 lbs.

* *Manuel de l'Ingénieur des Ponts et Chaussées*, 9me Fascicule; page 31.

TABLE No. 43.—RESISTANCE TO TRACTION ON COMMON ROADS.
(M. Debauve.)

Vehicles.	Diameter of the wheels.		Width of the tires.	Draft, in pounds per ton of the weight.		
				Uniform metalroad, at a walk, or at a trot.	Paved road.	
					at a walk.	at a trot.
	feet.	inches.	inches.	pounds.	pounds.	pounds.
Cart	6	0	2	72	47	63
Tumbril	6	1	3	69	46	—
	6	2½	4·4	67	39	—
	6	3	5·6	67	37	—
Waggon	6	5	6	65	40	—
Cabriolet	4	10¼	2	81	54	76
Jaunting-car	4	11	2	81	67	83
	2	10	2	81	67	83
Stage-coach	4	11	5.2	65	36	45
	3	1¼	5·2	65	36	45

M. Debauve makes the general deductions from this table, that the advantage of the pavement over the metalled road is considerable for waggons, is less for stage-coaches, and is nearly nothing for *voitures de luxe*, or the cabriolet and the jaunting-car. The effect of the tabulated values is, in summary, as follows:—

	RESISTANCE TO TRACTION	
VEHICLES.	On metalled roads.	On pavements.
Waggon	67 lb. per ton	38 lb. per ton.
Stage-coach	67 lb. ,,	45 lb. ,,
Cabriolet	81 lb. ,,	76 to 83 lb. ,,

M. Tresca tested the resistance of an omnibus on Loubat's system, adapted with wheels for running on a common road. The experiments were made on an inclined street in Paris, in good condition, having ascending gradients of 1 in 55, one part of which was paved, and another part was macadamised. The frictional resistance, after the gravitation on the incline was eliminated, was as follows:—

EXPERIMENTS AT BEDFORD.

Surface.	Gross weight. Tons.	Speed. Miles per hour.	Frictional resistance. Lbs. per ton.
Macadam	5·67	10·7	83
Pavement	5·67	10·1	66

The resistance to traction, of agricultural carts and waggons, was tested at Bedford in July, 1874, by means of a new horse-dynamometer designed by Messrs. Eastons and Anderson.* The first course was a piece of hard road rising 1 in 430; it was dry and in fair condition, largely made of gravel. The surface was, in many places, somewhat loose. The second course was along an arable field, growing oats, on a rising gradient of 1 in 1,000; it was very dry, and was harder than in average condition. The fore wheels of the waggons averaged 3 feet 3 inches, and the hind wheels 4 feet 9 inches in diameter; the width of the tyres was from 2½ to 4 inches. The weight, empty, averaged about a ton, and it was nearly equally divided between the fore and hind wheels. The cart wheels were, say, 4 feet 6 inches high, with tyres 3½ and 4 inches wide. The weight of the empty carts averaged 10 cwt. The loads were from 2 to 4 tons in a waggon, and 1 ton in a cart. The following results are deduced from the given data; the speeds averaging 2½ miles per hour: †—

Vehicle.	Draft on road.		Draft on field.	
	Total.	Reduced to a level. Per ton, gross.	Total.	Reduced to a level. Per ton, gross.
	Pounds.	Pounds.	Pounds.	Pounds.
Pair-horse waggon without springs	159	43·5	700	210
2 pair-horse waggons without springs	251	44.5	997	194
Pair-horse waggon with springs	133	34·7	710	210
1-horse cart, without springs	49·4	28	212	140

* See a report of the trials in *Engineering*, July 10, 1874, page 23.
† See *Manual of Rules, Tables, and Data for Mechanical Engineers*, 1877, page 962.

The results of Sir John Macneil's experiments on the tractive force required to draw a waggon on various kinds of road has already been given, page 52. He made other experiments on the traction of a stage-coach on a section of the Holyhead Road. The weight of the coach empty was 18 cwt.; and the weight of seven passengers in addition, allowing 1½ cwt. for each passenger, was 10½ cwt.; total weight, 28½ cwt. The first part of the following table has been arranged for various gradients and various speeds, from the statement calculated by the experimenter. The second part of the table is added to show the net frictional resistance after the elimination of the resistance of gravity:—

TABLE No. 44.—TRACTIVE FORCE REQUIRED TO DRAW A STAGE-COACH :—GROSS WEIGHT 1·42 TONS.

Gradient.	Total tractive force required.			Net frictional resistance per ton.		
	Speed in miles per hour.			Speed in miles per hour.		
	6	8	10	6	8	10
	lbs.	lbs.	lbs.	lbs.	lbs.	lbs.
1 in 20	268	296	318	76	96	112
1 in 26	213	219	225	63	68	72
1 in 30	165	196	200	41	63	66
1 in 40	160	166	172	56	61	65
1 in 600	111	120	128	72	78	81
Averages	—	—	—	62	73	79

The net frictional resistances at equal speeds, vary very much for different gradients, by some unexplained cause. They are a maximum for the steepest gradient, and a minimum for gradients of 1 in 30 and 1 in 40; for these they are less than for 1 in 600, which is nearly a level. The mode of action of the horses upon the carriage may have been an influential element. The averages show:—

FORMULA FOR RESISTANCE.

FOR A STAGE COACH:—
    At 6 miles per hour, 62 lbs. per ton frictional resistance.
    At 8   ,,   ,,   73 lbs.   ,,   ,,
    At 10 ,,   ,,   79 lbs.   ,,   ,,

With these may be associated, for the purpose of deducing a formula, the resistance of a waggon on a good road, given at page 52, at, say, 2½ miles per hour, 44 lbs. per ton, frictional resistance. Plotting the resistances per ton for the above four speeds, the following formula is deduced :—

Frictional Resistance to Traction of a Stage-coach.

$$R = 30 + 4v + \sqrt{10v}.$$

R = the frictional resistance to traction per ton.
V = the speed in miles per hour.

NOTE.—The formula is applicable for waggons at low speed. This formula is simpler than the formulas deduced by Sir John Macneil, page 53, and it is equally approximate and comprehensive.

M. Charié-Marsaines made observations of a general character on the performance of Flemish horses drawing loads upon the paved and the macadamised roads in the north of France, where the country is flat, and the loads are considerable. The results of his observations, published by M. Debauve, are here given in English measures, in Table No. 45.

TABLE No. 45.—PERFORMANCE OF HORSES ON ROADS IN FRANCE.
(M. Charié-Marsaines.)

Season of the year.	Description of road.	Weight per horse.	Speed in miles per hour.	Work done per hour, in tons drawn one mile.	Ratio of paved road to macadamised road.
Winter {	Pavement .	Tons. 1·306	Miles. 2·05	Ton-miles. 2·677	} 1·644 to 1
	Macadam .	·851	1·91	1·625	
Summer {	Pavement .	1·395	2·15	3·027	} 1·229 to 1
	Macadam .	1·141	2·16	2·464	

The table shows that there is a clear inferiority of performance in summer; and that the pavement is superior, even in winter, to the macadam in summer—

			Ratio.
Pavement, summer	3·027 tons drawn one mile,	1·86
„ winter	2·677 „ „	1·65
Macadam, summer	2·464 „ „	1·52
„ winter	1·625 „ „	1·00

The average daily work of a Flemish horse in the north of France is, on the same authority, equivalent to 21·82 tons drawn one mile, in winter; and to 27·28 tons drawn one mile, in summer; giving a mean for the year, of, say, 25 tons drawn one mile. The horses are powerful, and the roads are easy.

M. Charié-Marsaines observed that the harness lasted six years, over pavement; and only five years over macadam. Also, that the waggons lasted 7 years on the pavement, and 9 years or the macadam: whilst the horses lasted much less time on the macadam than on the pavement. The greater mortality on the macadam is ascribed to the inhaling of the fine silicious dust which rises from it.

It is stated by Mr. D. K. Clark,* that a good horse can draw a load of 1 ton at 2½ miles per hour, for from 10 to 12 hours a day; equivalent to (1 × 2½ × 10 =) 25 ton-miles per day. This is the same performance as is above given by M. Charié-Marsaines.

* *Manual of Rules, Tables, and Data for Mechanical Engineers*, 1877, page 720.

APPENDICES.

APPENDIX I.

ON ROLLING NEW-MADE ROADS.

By General Sir John F. Burgoyne, Bart.

[This paper, written in 1843, is valuable now, and is here reproduced from the fourth Edition of the present work.]

The importance of rolling roads, either newly constructed or when subjected to extensive repairs, seems never to have been duly appreciated.

Lines of any length of new-laid broken stone may be deemed nearly impracticable to ordinary traffic; the worst and most hilly old roads are always taken in preference to the new roads while in that state, although the latter may be much shorter, and with very improved levels.

At length the old road is shut up, carriages are forced to take the new, occasioning the greatest inconvenience and drawback to the intercourse for perhaps a year or more, a great wear and waste of the material, and a considerable expenditure in watching and maintenance, until the material, or what remains of it, shall be finally consolidated, and even then in a very imperfect form, unless great pains are taken with it.

The rolling is, in fact, effected, but in the most distressing and expensive manner, and by carriages and horses very ill adapted to it.

These evils may be entirely prevented, the road put at once into good working condition, and, certainly, a considerable expense eventually saved, by thorough systematic rolling; nor ought any road to be considered as *made* until that operation shall be completely effected.

Three reasons have probably operated to prevent this principle having been acknowledged and acted upon.

1. Because the traffic on the road will, sooner or later, do the work, thereby apparently reducing, in a small degree, the cost of the original construction or repair.

2. Because a roller is not usually at hand, and, from its weight and unmanageable character, it is most inconvenient and expensive to be removed from one place to another, so that in most cases one would have to be constructed for the purpose, and then be useless.

3. Much uncertainty, as yet, as to the best manner of operating, its efficacy, and the expense.

The first reason is founded completely on error; it is manifest that this manner of completing the road by the traffic is most inconvenient, and occasions enormous sacrifices by the parties using the road, and consequently a great loss to the public in general; nor can there be a doubt but that the *actual expenditure on the subsequent early maintenance of the road itself is greater than would be incurred by at once operating thoroughly with the roller.*

With regard to the second reason, there are many ways in which the objection can be greatly alleviated.

Although there is some justice in the third, and that the most perfect mode of proceeding is not yet perhaps understood, there is so much useful effect to be produced by any, that it is surprising that it has not been reduced to just principles by experiment, and generally adopted.

The practice of rolling has been rare, and almost entirely confined to gentlemen's demesnes, and occasionally to the macadamised roadways in some cities; but in the latter, it is believed, without the application of sufficient means for the purpose.

There are certain considerations which may serve as guides to arrive at just conclusions with regard to this proceeding.

1. A roller should not be too heavy in proportion to its bearing surface, or, instead of binding the material in the position and form laid down and desired, it will press it more or less into the substratum; much of the material will thus become useless, and it will be very troublesome to obtain the necessary resistance for the consolidation.

2. It must not be too light, or the effect will be too small ever to gain the object fully; or at any rate without an extent of operation that would be very costly or inconvenient.

It is believed that the ordinary rollers are too light, which may have thrown the practice into disrepute.

For the Dublin streets they have a roller of two contiguous cylinders, each of 4 feet diameter, and 1 foot 6 inches in width, making in all a bearing of 3 feet; it weighs 2 tons 3 cwt.; only two horses are attached to it, but the work is exceedingly heavy. It is applied to successive layers of material, in new formations, and about an inch of gravel is worked into the upper layer or surface. It is said to consolidate the roadway very effectually, but might probably be improved by adding to its weight.*

From other recorded trials, however, there is reason to believe, that a road roller should not be lighter than 28 cwt. for every 12 inches lineal of bearing on the road; that is, if 4 feet wide, that it should weigh 5 tons 12 cwt.; if 3 feet, 4 tons 4 cwt., &c.; and that it should only be applied to the upper surface of all.

A roller somewhat heavier than 28 cwt. per foot would be more effective, but it is better after that limit to gain the object rather by adding to the number of times passing over the surface than incur the inconveniences of the heavier machinery.

This is one very interesting point to prove, namely, the relative effects of light and heavy rollers, taking into account the number of turns required by each.

3. For effect, the wider a roller can be, the better, because the operation will be more quickly performed, and because, in proportion as it is narrow, will there be a tendency to force the broken stone laterally from under its action; but, as the weight must be in proportion to its bearing surface, the width must be limited to a degree that will prevent that weight being too unwieldy; a very narrow roller might also have a tendency to overturn. On the other hand, one that is very wide may take up too much room, if the road is open to traffic during the time of its use.

4. Horses should not be obliged to use very great exertions in drawing a roller, or the action of their feet will discompose the loose stones very inconveniently; therefore, as the draught is very heavy at first, and never very light up to the last of the operation, they should not have more than from ten to twelve cwt. each to draw at first, nor so much as a ton each at last.

5. It would be desirable not to put more than four horses to such a machine, because, as the number of horses is multiplied, it becomes more difficult to obtain a perfectly united effort from them; but on the above data a roller of four tons maximum weight might be too small for the best service, and as six horses may perhaps be applied without *much* inconvenience, it is proposed to give that number as a limit, and to allow 5 tons 12 cwt. as the maximum weight for the roller; this, at 28 cwt. per foot of bearing, would give it a width of four feet.

---

* A short street (Herbert Street), made in 1836, and then well rolled, has never required repair or new material since, up to 1843; it is a good street, but not entirely built on, nor a great thoroughfare.

From the Continent we have records of several trials that have been made of late years of the effect of rolling new-laid material on roads; although there are discrepancies in some of the particulars, there are many in which all agree; and in all the practice has been strongly recommended.

The one that seems to be the most practical is a roller described as first used in the Prussian provinces on the Rhine, and from thence introduced with some modification into a neighbouring district in France.

It consists of a cylinder of cast iron of about 4 feet 3½ inches wide and 4 feet 3½ inches diameter.* On the axle by means of iron stanchions is fixed a large wooden case of 6 feet 4½ inches long, 5 feet 8½ inches wide, and 1 foot 8 inches high, open at top.

This roller has a pole before and behind, in order to be able to draw it in either direction without turning; the hind pole is sometimes used to assist in guiding it. It has also a drag, by the pressure of a board on its face, in the manner used for French waggons.

The cylinder and other iron work weighs nearly 2 tons; the case and woodwork about 19 cwt., making the whole 2 tons 19 cwt.

The case will contain a weight of stone of 2 tons 19 cwt. when completely loaded; therefore the entire weight can be brought up to 5 tons 18 cwt.

Six *strong* horses worked it well.

It is passed over the entire surface of the road once or twice without any loading, weighing consequently nearly three tons, to obtain a first settlement of the loose material; then one or two turns with about 1½ tons loading, making 4½ tons; and then the last turns, making ten in all, with the full loading, when it becomes 5 tons 18 cwt.

Traversing 12 miles, it will thus completely roll about 3,000 square yards † in one day, or about a quarter of a mile of a road of 21 feet width.

All accounts agree as to the absolute necessity, and the best manner of applying some gravel, or other sharp, gritty, very fine stuff on the surface, during the operation, without which it will not be thoroughly bound.

The consolidation commences with the lower part, which is the first to get fixed and arranged; and when, after about six turns over the whole, the upper layers have become tolerably firm and well bedded, some sand or stone-dust, or, what is best of all, sharp gravel, is very lightly sprinkled over it by degrees at every successive rolling, *solely*

---

* These and other dimensions are necessarily in odd numbers, owing to reducing them from French measures and weights.

† Some of the calculations are not *strictly* in accordance with the data, because the data themselves are not given in minute fractional parts, and consequently the reduction of the results will show a difference; but it is very small and of no consequence in a general consideration of the matter.

for the purpose of filling up the interstices of the broken stone, and *not to cover it;* about 3 cubic yards in the whole per 100 square yards (equal to about an inch in thickness if spread over the whole surface) will be required. It is essential that this small stuff be not applied earlier, or it will get to the lower strata, and not only be wasted but injurious; the object is that it should penetrate for two or three inches only, to help to bind the *surface*.

Provided the upper interstices are filled, the less gravel used the better; therefore it is applied by little and little after each of the three or four last successive passages of the roller, and then only over the places where there are open joints.

After the work, if well done, is completed, it is stated that such is the effect, that the upper crust may be raised in cakes of six or seven square feet at a time, which could never be without the gravel.

The effect may be improved also by having the upper inch or two of stone finer than the rest, say to pass a ring of 1¼ inch or 1½ inch.

This work should be done in *wet* weather, or the material will require to be profusely watered artificially.

It will be better that it should not absolutely rain, unless *very lightly, when the gravel is applied* (although the stoning should be wet), as it will cause it to adhere to the roller, and even at times to bring up the broken stone with it. In frost it is of no use attempting to roll. The state of the material, as regards its being wet or dry, will have great influence in the success of the operation.

The form of the road will be best preserved by rolling from the two sides towards the middle, and not commencing along the latter.

The calculated expense of the work in France was—

For six horses and two drivers per day	£1 4 0
For six labourers attending on the road, assisting at the roller, levelling inequalities, spreading gravel, &c.	0 7 0
Total for 3,000 square yards	£1 11 0

Being about one penny for eight square yards, or one penny per running yard of road, twenty-four feet wide, and will amount to about £7 5s. per mile.

For Ireland, these prices would have to be increased, thus—

For six horses, with drivers, per day	£1 7 0
For six labourers, at 1s. 4d.	0 8 0
	£1 15 0

It is considered that a modification would be desirable in this foreign roller, by making it only four feet wide; its weight might then be, with its box for the additional loading, &c., about 2½ tons, which with an extra loading of 2 tons 2 cwt. would bring it to 5 tons 12 cwt. for its extreme weight, at 28 cwt. per foot.

Such a roller passing ten times over every part, and working twelve miles per day, would require five days, and the operation costs about £8 15s. per mile of road, of twenty-four feet wide completed.

The gravel ought to be considered as material, but in this case it is an addition to what would otherwise be applied; the cost therefore must be added.

Suppose it to amount to one shilling per cube yard, the expense, at thirty-six square yards per cube yard of gravel will be about £19 5s. per mile of road of twenty-four feet wide.

This would bring the whole to an amount of £28 per mile.

However perfect the rolling may be, there will be at the end a slight elasticity and yielding of the surface, which will only become quite firm and hard after some days' traffic, say from six to ten when tolerably frequented, during which its form and smoothness must be carefully attended to; add, therefore, £2 per mile for that extra work, and the cost will be £30 per mile.

The expense of the operation of the roller (independent of the gravel) is so small, that if the weight is under-estimated, so that the width of the roller should require to be reduced to three feet, thus adding one-fourth to that part of the outlay, or that it would require to be passed a greater number of times more than calculated, that increase would not be of essential importance on the gross amount.

If artificial watering should be necessary, that expense also must be added, but it would be small.

The subsequent wear of material, under proper care, will be most trifling; one French engineer states, that where the rolling in this manner has been successfully performed, there has never been a necessity for applying above one cubic yard of broken stone per 300 square yards of road in the next year; that in one instance only one cubic yard per 1,500 square yards was used, although on a road subject to the passage of 400 horses in draft per day; and on another road no fresh stone was laid for three years.

To make a more direct comparison, however, of expense, it may be assumed that a much greater diminution of thickness will take place in the consolidation by the traffic than if effected at once by the rolling, because the narrow wheels of ordinary carriages penetrate into the loose matter, and force the lower part of it partially into the subsoil. The displacement and grinding and crushing is also very great; whereas, in rolling, the entire is preserved and in its proper place; it may therefore not be too much to estimate, that if it require ten inches of loose material to bind into six inches by the ordinary process, as it probably would, eight inches well rolled would give the same; if so, the saving at once would be very great; thus, suppose the covering of one inch of stone to cost as much as two inches of gravel, that is, if the gravel is valued as above, at 1*s*. per cubic yard, that the stone be valued at 2*s*., then we have four times the cost of the gravel, which was stated to be £20 per mile, or £80, to set against the £30, estimated expense of rolling.

If the rolling effected a saving of only one inch of the broken stone, still the cost of that one inch would exceed that of the rolling, including the gravel.

This last calculation is given only as a proof of the saving, and not as recommending the reduction of the mass of material laid down to a minimum; on the contrary, as the rolling of the surface is a final measure, and requires no renewal until the road is worn to a minimum thickness, the most economical plan probably would be to apply a considerable degree of substance at once, enough to last some years, so as to reduce the number of periods when rolling would be necessary.

A roller, on what is here assumed to be the best construction, namely, of 4 feet bearing in one cylinder, and weighing 2½ tons, would not be very convenient to move from place to place for any considerable distances; an idea was therefore suggested of constructing one as a cart, on two wheels, the tires of which might be of 9, or 12, or 18 inches in width. Such a machine would weigh from one to two tons. The axle might be bent, and under the body, or might be straight, and pass through it, so that the loading of two or three tons might lie very low. The distance between the wheels should be a certain number of times the width of the tires; namely, if the latter are 9 inches, the distance asunder might be 3 feet, or 3 feet 9 inches; if 12 inches, they might be 3 or 4 feet; and if 18 inches, 3 feet.

Such a roller might be very convenient in many respects, but would be subject to two objections:

> 1st, The tendency to force the material laterally from under its pressure, by the little width of each roller.
>
> 2nd, The impossibility by successive passages of giving every part of the road precisely the same amount of rolling; some parts must have more turns than others.

In some few situations the very formation of the road may be made subservient to its rolling.

In the construction of a new road over the Carey mountain, in the County of Antrim, material for stoning the road was quarried in several parts of the mountain up to its summit.

Some carts were made with wheels of four-inch tires, and the laying of the broken stone being commenced close to the quarry, the work was carried on from each quarry down hill, the loaded carts being taken over the new-laid material, working systematically over the entire width of the road, and discharged below, returning up the hill light. By the time the work was completed the road had acquired in this way a considerable degree of consolidation without extra labour.

A roller of the weight of five or six tons may be worked up inclines of one in twenty by increasing the number of horses, but not steeper; if at all exceeding one in thirty it would probably be better to apply the roller in its lightest state, and increase the number of passages.

It is very desirable to complete rapidly what is once begun, but it is attended with the disadvantage of taking up short lengths at a time, which leads to the occasion for turning the roller very frequently, a manœuvre that is particularly inconvenient.

Although certain dimensions and weights are suggested to be likely to prove the most efficient, any other kinds that happen to be in possession might be tried and adapted to the above principles, which will usually require weight to be added with the successive rollings; this may be done in various ways according to circumstances and situations; the most simple, but most cumbrous and troublesome of application, will be a large case on the roller for loading with stone.

In or near towns, iron weights might be used instead of stone, partly on or suspended to the axle, within the cylinder, or in a case outside, which might be then much smaller, and the weight be more compact and more easily shifted; or for use in a town, when the most efficient dimensions and weights were ascertained, rollers might be prepared of two or three qualities, that is, all of the same extent of bearing, but of cylinders of different weights, from the lightest to the heaviest, and brought in succession on to the work.

APPENDIX II.

EXTRACTS FROM "REPORT ON THE ECONOMY OF ROAD-MAINTENANCE AND HORSE-DRAFT THROUGH STEAM ROAD-ROLLING, WITH SPECIAL REFERENCE TO THE METROPOLIS."

(Printed by Order of the Metropolitan Board of Works.)
BY FREDERICK A. PAGET, C.E. 1870.

THE SPECIAL SUITABLENESS OF ROAD-ROLLING TO A MOIST CLIMATE.

There is a point much in favour of an extended use of road-rolling in London and England generally; making it probable that there is no other country in the world from which such good effects are to be expected from steam road-rolling. The climate of England, from its humidity—and the best judges, Macadam amongst others, consider the presence of moisture as extremely favourable to road-making—and its freedom from either long frosts or long droughts, is peculiarly favourable to the employment of rolling. In France and Prussia the rollers are used by preference in rainy weather, in order to save watering, as wet is essential for consolidating the metal. From want of experience in road-rolling some English surveyors object to the application of sand, which it is necessary to use—of course in not too large quantities—with rolling, in order to prevent the crushing of the metal. Upon the Holyhead roads, consolidated in the ordinary way, Telford always applied clean gravel to the new metal; and this is now done in France, Prussia, and on all roads where road-rolling is understood and practised. The binding of rolled roads in Prussia by means of the sand, is so strong that united blocks 18 inches wide can often be removed.

According to the official Hanoverian "special instructions" on road-rolling, when rolling the covering of the road, the binding or sand is only to be applied by the time that about two-thirds of the operation is completed. The amount of sand must be only about one-sixth of that of the metalling. The watering should be very carefully carried out, so that the bottoming be softened as little as possible.

THE APPLICATION OF THE STEAM ROLLER TO THE FORMATION OF THE FOUNDATIONS OF NEW ROADS.

France.—Except with very heavy traffic, or with a very soft substratum of the road, the *Ingénieurs des Ponts et Chaussées* do not use paved, and very seldom concrete, foundations or bottoming. They compress the bottom by rolling; and then roll down upon it successive layers of road material about four inches thick.

Germany.—When rolling road bottomings, the German engineers do not take the roller over the sides, but rather lightly compress the crown of the road foundation. According to the official Hanoverian " special instructions " about rolling the bottomings of roads, this must be done with a light roller, and must not be carried out till the bottom is quite hard. No water must be used, and the operation should be carried out in dry weather, especially with a clayey soil.

Liverpool.—The thirty-ton steam road-roller, by Messrs. Aveling and Porter, was set to work in Liverpool in October, 1867. As we have already remarked, Mr. James Newlands, the borough engineer of Liverpool, has not yet been able, in so short a time for such a purpose, to determine the actual saving in maintenance effected by it. He writes us, however, that " besides its advantage in making a newly-coated macadam road perfectly smooth in a single night, it is of no less utility in forming the foundations of new roads. Formerly the traffic had to be turned on these foundations to consolidate and render them fit to receive the protective coatings of paving or macadam respectively—a work which took from three to six months, according to the locality. Now, when the foundation is laid, it can be rendered fit for paving and macadam in a day or two." There are upwards of forty miles of macadamised roads in the borough of Liverpool.

SUBSIDIARY APPLICATIONS OF THE STEAM ROAD-ROLLER.

Gravel Roads.—As regards the application of the steam-roller to gravel roads, of which areas still greater than those of granite and flint roads are in existence, but little has yet been done except in parks in this direction, and certainly nothing has yet been published as to the economical results. Macadam wrote, some fifty years ago, " In the neighbourhood of London, the roads are formed of gravel," " the component parts" of which " are round, and want the angular points of contact, by which broken stone unites, and forms a solid body;" and large areas are still laid down on the outskirts of the metropolis. The only surveyor who has spoken from experience as to the economical value of rolling gravel roads is Mr. Howell, of St. James's. But, as has been well observed to us by Messrs. Amies and Barford, " If a garden-roller is useful in a degree, surely a much heavier one must be still more efficient for light vehicle traffic."

Paved Streets.—In Paris, successful, but partial, experiments have been made with the steam road-roller as a substitute for the paviour's rammer.

Heavy rollers are very extensively applied in Hanover to pavements; and the official Hanoverian instructions for keeping up the roads contain several directions as to the time and way in which the roller is to be applied. Experience has shown that the rollers should be as heavy as possible; that they should be applied to any unevenness of considerable area; that pavements are best rolled after the breaking up of frost; and that the surfaces should first be swept, and then the rolling be continued until the stones no more give way. Herr von Kaven, in his "Lectures on the Engineering Sciences," gives elaborate figures, based on some experiments conducted in Hanover in 1863, showing the economy of this method of laying down and keeping pavement in repair.

It is clear that rolling must also be peculiarly applicable to the construction of foundations for paving setts.

Picking up Roads by Steam.—The steam road-roller, though still a comparatively new implement, has thus already made itself useful in a variety of ways, and has adapted itself to several important functions in road-making and repairing, besides that of rolling only. With respect to picking up, or "lifting" the macadam of old roads, before laying down a fresh coating of metal, there can be no doubt that a great economy of manual labour can be effected in this direction. Mr. Aveling's apparatus for "lifting" roads simply consists of a series of spikes bolted on plates let into holes in the periphery of the roller. A machine on this principle, the invention of Mr. Browse, the general surveyor of the Metropolis Roads Office, has long been used by Messrs. Mowlem, Burt, and Freeman, for picking up roads in the metropolis. It simply consists of a heavily-laden cart, having its wheels armed with spikes. At Manchester, one of Aveling and Porter's engines picked up an area of 2,048 square yards in three hours forty minutes; and the cross-picking was afterwards done by hand labour, only equal to that of one man working sixty hours.

This application of steam will no doubt lead to very great economies on many country and other roads that have been long wastefully managed. What Macadam wrote still applies in many cases. "Generally the roads of the kingdom contain a supply of material sufficient for their use for several years, if they were properly lifted and applied" —with the difference in favour of the present time that much of this metal has been previously broken. On any soft, ill-drained foundation, successive supplies are often driven in by the superincumbent traffic, the intermediate mud working its way up. Some road surveyors consider that many old roads, on which there is often as much as two feet of metal, might be mended by the steam-roller without using any new metalling. They could be picked up and then steam-rolled, with an excellent new surface as a result.

Breaking and Conveying Metalling.—It is obvious that the steam road-roller could be easily applied, not merely to driving stone-breaking machines, but also conveying the metalling to the spots where required.

SUBSIDIARY ADVANTAGES OF THE STEAM ROAD-ROLLER.

Saving in Sweeping and Scavengering.—In all these instances, it is quite clear that any saving in road metal must be accompanied, as in the case of the substitution of pavement for macadam, though, of course, to a slighter extent, by a proportionate saving in the expenses of scavengering and removal of detritus.

Diminution in Paved Areas.—It is also clear that its well-considered and properly carried-out adoption in our large towns must tend, as in Paris, to narrow the encroachments of the uncomfortable, if cheap, paving stones. The main advantage, in certain situations, of pavements is in the diminution of cost of maintenance. At speeds higher than at a walk, General Morin's experiments prove that the resistance to draught on pavements increases, which is not the case with a well-metalled road in a dry state.

RÉSUMÉ OF THE MAIN ADVANTAGES OF STEAM ROAD-ROLLING.

Briefly, the three principal advantages obtained by steam road-rolling over horse-rolling are, as observed in the French Official Report on the Exhibits at the late Paris Exhibition, when alluding to the Ballaison steam-roller, worked in Paris by Gellerat and Co., "saving of expense, saving in time from rapidity of execution, and a better quality of work." All that has been said with respect to the great diminution in draught by horse-rolling applies with much greater force to rolling by steam. But even if steam-rolled roads cost more instead of at least fifty per cent. less in maintenance than traffic-rolled roads, there would be good reasons for adopting steam-rolling, as the public is well served under the first circumstance, and very badly served under the second; so that, even if the expenses were greater, there would be a balance of advantage and expense. As a correspondent, "H. R.," in 1868, observed in a letter to the *Times*: "When one considers the injury done to carriages and horses, and the delays caused by the practice of leaving the stones to be ground in by the carriage-wheels, one feels that even if rolling in the stones would be more costly to the parishes, the pecuniary gain to the public would be very considerable." In the words of Mr. Robert Mallet, the well-known engineer and author, writing in 1866, "there is still room for very great improvement in the laying and repairs of the macadamised streets and roads of the metropolis, although the practice has of late years in some respects improved. The admirable practice of Paris, where, with far worse material than is at our command, the results obtained are so much better, should be in part our school and example.

"We have tardily adopted from thence the steam road-roller; let its use be extended, and let us also adopt the admirable methods of cleansing, watering, and repairing of the French." The question is, as Mr. Tomkins, the surveyor of the roads in the district of St. George's Hanover Square, very neatly defines it, whether the roads shall be consolidated "by the traffic or for the traffic."

ECONOMICAL RESULTS OF THE GENERAL USE OF STEAM ROAD-ROLLING IN THE METROPOLIS.

Remarks on the Table.—The annexed table* is almost sufficiently explanatory of itself. It has been made as complete as has been possible; and when we state that, without counting the Parliamentary Returns, not fewer than forty-four different authorities have been applied to for returns, and that probably each different authority keeps its books on a different plan, some notion of the labour undergone in compiling the table will have been formed. It will be seen that the thirty-eight Metropolitan Vestries and District Boards of Works appear in the same order as usual in such returns; the Vestries being separated from the Districts. If it had been possible, we should have preferred a different arrangement, based on other considerations. The great roads of London, and their minor feeders, radiate from the City as a centre, something like the threads of a completed spider's web. As we leave the centre, the main roads necessarily diverge farther apart, until lost in the surrounding country. If the metropolitan area were divided into small regular squares, or hexagons, of equal areas, it would be possible, starting from the City as a centre, to pretty closely classify portions of the divisions, now coming under the headings of parishes and districts, into four principal zones. Within a circle of about two miles radius from St. Paul's, enclosing an area of some thirteen square miles, we should find that all the streets are paved. Another circle, or perhaps rather an ellipse, with the Thames from the Tower to Somerset House as its major axis, would enclose the principal Guernsey granite areas; farther out we should find, still of course dealing with the generality of minor roads, a zone mainly composed of flint roads; until, at the farthest distances from the City, and of course in regions of lesser traffic, we reached the outermost zone of gravel or pebble roads. It is of course at once seen that, as the traffic increases towards the centre of the web, the roads have to be made stronger and

---

* The table here referred to is given in abstract at page 152, *ante.*—D.K.C.

stronger, until a centre is reached entirely paved with the hardest granite.

We must note that the term "macadamised" is by some London surveyors strictly confined to a road laid with granite; by others it is extended to flint roads, which again are, by some surveyors, put under the category of gravel roads. We have tried in the table to separate as much as possible the granite from the flint and gravel roads.

It will be observed how the average width of the streets and roads increases in the wealthier parishes, and gradually diminishes in the poorer and older ones. In St. James's, Westminster, and Battersea, the roads are, on the average, 50 feet wide; in Paddington we have an average width of 45 feet; in St. George's, Hanover Square, 40 feet; and in St. George's in the East only 18 feet.

The Present Extent of the Roads in the Metropolis.—Mr. George Vulliamy, architect of the Metropolitan Board of Works, in his "Memorandum as to the Means of providing Fire-plugs in the Metropolis," assumes the length of streets in London at from 2,000 to 2,500 miles. This rough estimate includes the paved roads. Our table, based on thirty-six different returns, from as many parish and district surveyors, gives more than 1,100 miles of macadamised roads in the metropolitan limits, with an estimated total area of nearly twenty-three millions of square yards. These totals are really far below, rather than over, the mark, as they do not take into account the areas of private roads, of which there exist a great number, especially at the outskirts, such as towards Battersea, Lewisham, Islington, Hackney, and other districts. Besides, in the few cases, such as Marylebone and others, in which incomplete returns of lengths or areas could be obtained, the figures given have been estimated low.

These totals also exclude the gravel footpaths, the areas of which are very considerable in many parishes.

The public and private parks are also not included. In the former, Hyde and St. James's Parks alone contain eight miles of roadway under Government supervision. With Regent's Park, Battersea, Victoria and the other Parks, at the very least thirty miles of road could be added.

It will be observed on looking at any map, that the metropolitan area of parishes has grown into the shape of a huge bunch of grapes from the Thames as the stalk of the whole bunch. Between the fringes of its irregular contour are great areas of inhabited country, often as much covered with roads as the adjacent metropolitan parishes proper. There are thus immense areas of roads on the outskirts. For instance, Mr. Browse, surveyor of the Metropolis Roads Office, has not less than seventy miles of macadamised turnpike roads, beyond the metropolitan area, but still in Middlesex. These seventy miles have an average width of 40 feet, without reckoning the footpaths.

APPENDICES. 315

Thus, taking the districts of Plumstead and Lewisham, the farthest from the centre of London, as giving a measure, in the ratio of the area of their roads to acreage, of the proportion of roads on the outskirts, we should find that, within a radius of only twelve miles from Charing Cross, there are at least forty millions of square miles of macadamised roads.

The Future Extent of the Roads of the Metropolis.—As might naturally be expected, considerable areas of roads are being laid down every year in the metropolis, and especially in the outlying metropolitan districts. In many districts, such as, for instance, Wandsworth, at least two or three miles are laid down annually.

The total population of 1861 was three times that of 1801, having trebled itself in sixty years. From the figures giving the population of London at the end of every decade since 1801, it can be seen that, since the beginning of the present century, the metropolitan population has doubled itself in about forty years. In fact, the Registrar-General estimates that the metropolitan population, which in 1866 was more than 3,000,000, will, at the present rate of increase, in forty years be doubled, or will rise to 6,000,000. Should the population double itself in the next forty years, the mean annual rate of increase, calculating from 1865, during that period, will be about 75,000. After making every allowance for the "tendency to build houses much higher than was formerly the custom," and without taking into account the opposing and corrective tendency to lay out wider streets than formerly, Mr. Haywood estimates that "at the expiration of thirty-nine years hence, perhaps forty or fifty square miles of open country will be covered, more or less closely, with houses, for the additional three millions of inhabitants which will then exist. But," he goes on to say, "there are other causes which have arisen of late years tending largely to disperse and radiate that part of the existing population which is above the operative classes, the principal agency being the facilities for transit offered by railways. The tendency of that class, undoubtedly, is to seek cheaper residences and a purer atmosphere, and consequently to encroach still further upon the open area surrounding the metropolis, so that probably sixty square miles of open country, if not a considerably larger area, will be covered and occupied by the time the population reaches six millions." This would give an area one-half more than at present.

The minimum Annual Saving to London Ratepayers derivable from the use of Steam Road-rolling.—After taking the most minute precautions against error, we feel some confidence that our figures of economy through road-rolling are much below the mark. We see by the table that there are at least 1,126 miles of macadamised road in the thirty-nine metropolitan parishes and districts, and that their total area is at least 21,562,605 square yards. Of course, the true measure of the

P 2

work to be done would be a figure giving the exact cubic contents of exactly the same kind of metalling consumed in a given period of time over the given area. It is impossible to obtain this, though we know that the surfaces of many London roads have to be renewed several times a year. We find, however, as can be seen in the table, that there were expended on the roads and streets in the London parishes and districts during the year 1866-7 £714,662; and during 1867-8, £781,003. From a number of returns forwarded to us by different London surveyors, we estimate that the average annual cost of maintenance, exclusive of cleansing and watering, of the London macadamised roads, is £280,750, or at the rate of, in round figures, £250 per mile.

We have seen, from not less than seven estimates, that the saving in maintenance through horse road-rolling can be estimated at 40 per cent., and that the French, after nearly ten years' experience in Paris, consider that they save 50 per cent. in maintenance *over and above* horse-rolling. Therefore, even if we assume that only 50 per cent. is saved by steam-rolling, we have a gain of more than £140,000 per annum, without reckoning the diminution of scavengering and watering, and many other sources of economy.

It will be observed that this sum is less than one-fifth of the grand total of that spent in 1867-8 on all the charges for the London roads; and that, capitalised at three per cent., it represents more than four and a half millions.

The Classes directly interested in the Application of Steam Road-rolling.—As every one uses roads, and is directly or indirectly a ratepayer, every one may be said to be more or less interested. This of course applies more directly to every owner of a horse, for the reasons we have already given at length, under the heading of saving in draught; but still to every user whatsoever of the road. As a London local paper, the "Parochial Critic," observes, "an uncovered part is seized upon by every humane driver, and quite a contest takes place for possession of the smooth portion. When all is covered by loose sharp stones the drivers and horses yield themselves to their fate, and drag through the punishment as best they can. A return of the number and character of accidents occasioned by the present system of road-making for the last seven years would surprise the ratepayers."

As usual in such cases, although the complaints of the wealthy are heard more loudly, the poor probably suffer most. It is said that, in some parishes, the habit is to lay down the loose stones in October, and to leave them to be consolidated by the passing traffic before the rich inhabitants return to town in February. The work of rolling is thus thrown on draught-carts, the vehicles of the poorer classes, and especially the omnibus—the poor man's own conveyance. The diminution in the work of traction produced by road-rolling certainly ought to direct the attention of the omnibus companies, now serving a passenger traffic in which they have to compete with the metropolitan railways, and soon with the tramways—the proprietors of railway vans, such as those of Messrs. Pickford and Chaplin and Horne, and of the railway coal-carts—to this mode of saving wear and tear of horseflesh and vehicles. Any road whatever is neither more nor less than an apparatus for saving draught; and the only advantage of a railroad is that it has so much less draught on its rails that it economises the work of traction. The roads of a city are in fact the permanent way and fixed plant of its means of locomotion.

The Amount of Capital invested in Horses and Vehicles in the Metropolis.—In order to obtain some further measure of the extent of the classes directly interested in the application of steam road-rolling, we will form an estimate of what an engineer would term the rolling-stock of the London roads. According to recent fiscal arrangements, all horses whatsoever are taxed, and also all vehicles, with the sole exception of trade carts for burthens, on which are painted the name and address of the proprietor. This gives an easy means of getting at the number of horses and, approximately, the number of vehicles. The Honourable Commissioners of Inland Revenue have obligingly furnished us with figures, giving the number of horses, and of taxed vehicles, within a circle of a radius of $4\frac{1}{2}$ miles from Charing Cross, and hence covering the greater part of the metropolitan area. There are thus 71,903 horses in all, and 24,095 taxed carriages, within this space.

On account of the untaxed trade carts, we may safely raise the number of vehicles to 30,000. In the absence of any exact data, there may probably be different opinions as to the relative numbers of the different kinds; but the following rough estimate will not be very far from exactitude, especially as we have verified the numbers we have estimated for the cabs and omnibuses by inquiry at Scotland Yard. Of light carts, vans, gigs, phaetons, 15,000, which, at the price of £30 each, would give a sum of £450,000; cabs, 7,000 at £35 each, give £245,000; omnibuses, 2,000 at £60, £160,000; heavy carts and waggons, 3,500 at £50 each, £175,000; carriages, broughams, &c., 2,500 at £120 each, £300,000. This would afford a total of £1,290,000 invested in vehicles; and harness may be estimated at 40,000 sets, at the rate of £4 each, or £160,000. The depreciation of this capital is very heavy, as experts estimate that vehicles, the first prices of which are £20 to £150, cost from £4 to £15 per annum in repairs; and the wear and tear of harness is always very considerable.

Each of the 15,000 light cart, gig, phaeton, or van horses we will take to require one horse, of the value of £20, or a total value of £300,000. Each of the £7,000 harder worked cabs would require two horses, as a horse must be in stable at least three days a week; but their average value can only be taken as £15, or a total of £225,000. As few omnibus horses do more than one journey a day, and even then often rest one or two days a week, we probably have 30,000 horses to the 2,000 omnibuses; and being of a somewhat superior kind, they must be estimated at the rate of £30 each, or a total of £900,000. Heavy cart or waggon horses, 7,000 at £60 each, and two per cart, will give £420,000; and carriage and brougham horses, two per vehicle, 5,000 at £100 each, will employ the sum of £500,000. On account of their inconsiderable number, especially in the centre of London—for we are only considering the area within a radius of 4½ miles from Charing Cross—we have left out the riding-horses. The total capital thus invested in horse-flesh we may estimate at £2,345,000; and, adding to this £1,290,000 for vehicles, and £160,000 for harness, we have a total of £3,735,000, or of nearly four millions of pounds sterling, invested in the rolling-stock of the metropolitan highways.

THE HISTORY OF HORSE ROAD-ROLLING.

The first recorded allusion to road-rolling seems to have been made in the letters patent granted in 1619 to a certain John Shotbolte, who, without enrolling any specification, speaks of using "land stearnes, scowrers, trundlers, and other strong and massy engines," "in the making and repairing of highways and roads." In Partington's "British Cyclopædia of the Arts and Sciences" (1835) we find it stated that it was soon after the year 1700 that a part of the charge of repairing roads was taken off the parishes and levied on the traveller by means of turnpike-gates. It was, nevertheless, a complaint that the roads were little, if at all, improved by the expenditure of the money so raised. This complaint is energetically advanced in "a dissertation concerning the present state of the high-roads of England, especially of that near London, wherein is proposed a new method of repairing and maintaining them," read before the Royal Society in the winter of 1736-7 by Robert Phillips, and printed in a small separate volume. "The author's great object is to recommend washing the roads by a constant stream, if possible;" at any rate washing the materials—in which he partly anticipates Macadam—which they are composed. Phillips strongly remonstrates against "the practice of laying down large heaps of unprepared gravel to be gradually consolidated into a harder mass, at the expense of the intolerable labour of the poor animals that are obliged to grind it down." His complaint has been re-echoed in London for more than one hundred and thirty years, but with little practical results.

The first proposal of a road-roller, on a sound and scientific basis was made by a French Royal Engineer of Roads; and already, 1787, M. de Cessart, then *Inspecteur Général des Ponts et Chaussées* recommended a cast-iron roller for rolling down newly metalled roads. He chose it 8 feet long, 3 feet in diameter, and of a weight of 7,000 old French pounds. Very soundly he compared its action on broken stone roads to that of a pavior's hammer on paving stones, and altogether his description is what might be expected from a scientific engineer. The *Assemblée des Ponts et Chaussées* strongly advised a careful trial; but the disturbed political state of France, then in the throes of her first Revolution, evidently had the effect of preventing the continuation of the experiment. De Cessart's description is republished in the volume for 1844 of the *Annales des Ponts et Chaussées*. Philip Hutchinson Clay patented and specified in England, in 1817, " a large roller which, with the carriage over it, empty or filled with gravel or other material, may be made to press upon the road any weight from 6 tons to 20 tons, and by being used frequently, will press the moisture to the surface, and admit the sun and rain to act upon it, at the same time pressing down the loose material and making the road more even." That Clay understood the subject is evident from his using " a gravel cart," " fixed upon a roller, to carry gravel or other material used in the repairs of roads, and deposit it where necessary, the great advantage being that it improves the roads it passes over, and will be particularly useful in districts where material is scarce, whereas the present mode of carrying it in carts with narrow wheels destroys the road it passes over." This plan was afterwards carried out in Ireland by Sir John Burgoyne. John Biddle patented, in 1825, " a combination of machinery for making, repairing, and cleansing roads, consisting of scrapers combined with rollers." The road-roller thus seems to have been used by several English inventors; and it is acknowledged by the French that a sight of some at work in London revived the attention of Continental engineers to their use. A road-roller was tried as far back as 1826 on the road near Göttingen from Hanover to Cassel; but the experiments did not succeed at the time, and it was only between 1837 and 1840 that Herr E. Bokelberg, now *Wegbaurath* (literally " Road-making Councillor "), practically introduced the process into Hanover. The first roller adopted in actual practice in France appears to have been one by M. Polonceau, the celebrated builder of the Carrousel Bridge, in 1829. This roller is said to have been built up of wooden lagging tied together with iron hoops, and weighted with stones, being, as to construction, long behind that of De Cessart. Wooden rollers, filled with stones, sand, or clay, were in use in France, about thirty years ago, in the department of the Marne et Loire; and one of this kind is illustrated in a volume for 1840 of the *Annales des Ponts et Chaussées*. Rollers made of solid stone and granite were also used at the beginning. Some French writers state that the horse-roller was first re-introduced from England into France, in 1833, by M. Fortin. Dr. Rühlmann, of Hanover, unknown to ourselves, arrived some years ago at the same conclusion that road-rollers first began to appear in actual practice in 1830, or thereabouts; and that they were introduced into Hanover in 1837, or after their appearance in France. As we know, the Hanoverian administration of the roads has the reputation of being the very best in all Germany. The year 1830 is about the date given by Mr. Vignoles as that of the general and official adoption by the French of Macadam's system of roads, of which they at once perceived that rolling is the indispensable concomitant. They hold that an unrolled macadamised road is only half finished; and the centralization of the French, Hanoverian, and Prussian administrations of the roads has facilitated the universal combination of rolling with the system of macadamisation. In 1837, Mr. (now Sir John) Macneil patented the use of a kind of road consisting of pieces of cast or wrought iron mixed with broken stone and gravel. It was to be consolidated by horse-rolling in preference to wheel traffic. That, however, road-rolling was not much known in England by 1838, appears from an examination of Sir Henry Parnell's celebrated work on roads, the second edition of which was published in that year. The Right Honourable Sir Henry Parnell, Bart., was constituted the Government Commissioner of the Holyhead roads, laid out by Telford; and the plans, specifications, and contracts drawn by Telford are printed in the body of the work. No mention is made of road-rolling—as little as by Macadam himself, in the sixth edition of his work, published in 1822. It is well known that Sir John Loudon Macadam rather directed his attention to the construction of new roads and to restoring bad ones than to the maintenance proper of good roads. Some of his principles are no doubt applicable to maintenance as well as to making and repairing, but there is a good distance from that to a complete method. The first recommendation in the English language of horse road-rolling, as a measure of economy, was published by Colonel (now Field-Marshal) Sir John F. Burgoyne, R.E., in 1843, when Chairman of the Board of Works in Ireland. He was one of the first engineers who used it in their own work, and certainly the first in England to scientifically recommend it, not as a refinement, but as a necessity. Before 1840, as appears from the *Verhandlungen des Vereins zur Beförderung des Gewerbfleisses in Preussen*, road-rolling was officially and universally introduced upon the Prussian roads. There can be little doubt that sooner or later all the rest of Europe will imitate Prussia in this, the only logical and truly economical mode of road maintenance, just as the Prussian authorities are being tardily copied in their first use of breechloaders, in their army administration, compulsory education, preliminary examination of patents for inventions, the stringent management of their civil services, and other matters.

A great number of different constructions of horse-rollers are in use, designed with regard to facility in turning at the ends, facility for increasing or diminishing the load on the roller, and ease of draught. The most elegant of these contrivances is the roller ballasted with water; obviously affording a very easy means for lightening or increasing the load. It is generally believed in England to be a home invention, though really first applied and described in 1853, by a Government Engineer, G. Nell, employed in Prussian Silesia.

THE HISTORY OF STEAM ROAD-ROLLING.

In the presence of the settled employment in France of horse-rollers, it was natural to expect that French inventive talent should first seek for the usual substitution, in an implement in much requisition, of steam for horses. The limitation of the areas paved with "setts" in Paris and other large cities, said to be ordered by Napoleon III. as a timely precaution against street barricades, has also greatly favoured the employment of the steam-roller by raising the demand for a very high and durable class of macadam, especially for the centres of Parisian traffic. The main impulse towards the use of the steam-roller in France was, however, the extraordinary economy in maintenance found to be derived from horse, and *à fortiori* to be expected from steam, road-rolling. The saving to the public in wear and tear of horse-flesh, vehicles, and harness, however great, could scarcely have so much stimulated the road authorities. The publication more than a quarter of a century ago of these economical results of horse-rolling has had little or no effect on *our* road management; the yet greater advantages obtained in France from steam road-rolling may, however, possibly do more.

The first patent for a steam road-roller was taken out in France, at the beginning of 1859, by M. Louis Lemoine, of Bordeaux. His roller was not patented in England, nor ever fully described in any publication. At the meeting on the 13th of February, 1861, of the *Société d'Encouragement*, the well-known engineer, M. Combes, is recorded to have described—though his description is not given in the *Bulletin* of the Society—Lemoine's roller, which he had seen at work on the Bordeaux roads. M. Lemoine was then an employé of the Bordeaux municipality. This steam-roller appears to have consisted of one main roller for doing the work; the engine, weighing in all only about ten tons, being steered by a pair of side wheels, the axles of which, by means of gearing, could be set at varying angles to the main roller. In 1860 M. Lemoine's roller was tried on the road of the *Bois de Boulogne*. Ballaison's steam-roller, patented in France in August, 1860, and now in use by Gellerat and Co., the Paris Steam Road-rolling Company, was also there tried in August of the following year. In 1862 experiments were carried out with both these steam-rollers by the French engineers of the State, M.M. Darcel and Labry, who gave an account of them in an unpublished report of May of that year. They expressed strong opinions in favour of steam-rolling in general, giving the preference to the Ballaison roller. An extract from this report was printed in M. Homberg's *Notice sur les voies empierrées et asphaltées de Paris* (1865). They observe that the only possible apparent objection to steam-rollers is their frightening the horses. We understand from M. Lemoine that since these trials he has made several different forms of rollers, now at work in Bordeaux and its environs. By this time the system has firmly established itself in Paris and elsewhere in France by a success extending over more than seven years.

The results obtained in Paris during the couple of years after 1861 attracted little or no attention in England. The main reason for this is to be found in the slight importance attached by most of our road authorities to rolling. It is not regarded as a necessity but as a luxury. Quite independently of the French inventors, Mr. W. Clark, chief engineer to the municipality of Calcutta, in 1863 conceived the idea of a self-propelling steam road-rolling engine, with its weight uniformly distributed over the whole width of the rollers. The different English traction engine-makers to whom Mr. Clark applied were unwilling to make designs and patterns for a single implement; and at last Mr. Clark having consulted Mr. W. F. Batho, of Birmingham, the engineering manager of Mr. Josiah Mason's vast undertakings, and well known in the profession for his original constructive talent, an engine on Mr. Batho's designs was made by Mr. Worsdell of that town and sent out to Calcutta, where it has done good service. This design, patented in 1863 by Messrs. Clark and Batho, was therefore that of the first steam road-roller ever tried or patented in Great Britain. The chief features of Clark and Batho's patent are the use of three sets of rollers, two in front acting as drivers, while the third is set up in a turn-table, being adjustable so as to steer the engine, at the same time overlapping the space between the two outside drivers. At about this time Mr. Batho went to much trouble in trying to prevail upon the road authorities of many of the larger towns to adopt steam-rolling, but without the least success. In a history of the subject he must, however, figure as the first Englishman who fully comprehended the national importance of steam road-rolling; and we understand that, although, as we have stated, the first experiments were made in France, yet his steam road-roller was the first thoroughly successful implement of the kind. About a year afterwards, Ballaison's steam-roller appeared on the English patent lists as a communication from Messrs. E. Gellerat and Co., of Paris; and in May, 1865, that Company concluded a six years' contract with the Administration of Paris for rolling the roads of that capital. In the meantime the roller made by Mr. Batho for Calcutta had been at work there since 1864, more than repaying its entire first cost. The Bombay authoritics accordingly instructed their agent to order one of Clark and Batho's engines, through Captain Trever, R.E., who, however, preferred to employ Messrs. Moreland and Son, of London. These gentlemen accordingly made two engines which were sent out to Bombay. In 1866, Messrs. Eastons, Amos, and Anderson undertook some contracts for rolling roads. Mr. Anderson informs us that the firm "happened to have one of Aveling and Porter's 12-horse traction-engines, and it occurred to them to harness it to a 10-ton roller," 6 feet in diameter and length. They "commenced operations in 1866 by rolling the new roads of the Belvedere Estate," Belvedere, near Erith. The firm "subsequently made a contract with the First Commissioner of Works to roll the roads in Hyde Park," and, "during the autumn of 1866 and the early part of 1867" the engine and roller were kept in that park, though the Government officials were anything but favourable to its employment. It again illustrates the little attention paid in England to what is being done abroad, that, though the city of Paris had, nearly a year before, concluded a six years' contract with the Paris Steam Road-rolling Company, yet this eminent firm employed a certainly ingenious, but comparatively inefficient, combination for the purpose. It was soon found that the wheels of the traction-engine left deep ruts that were not made good by the roller, and that much time was lost in turning; but that the dead load of the boiler and engine—useless in the traction-engine—could be made to do good work in rolling. What with these observations, and the reports probably more or less current about the success of the Paris and the Calcutta rollers, Messrs. Aveling and Porter determined to adapt their form of traction-engines to this purpose. The result is a combination of their simple and efficient form of traction-engine with the arrangement of rollers and turn-table patented in 1863 by Messrs. Clark and Batho. Towards the end of September, 1867, Messrs. Aveling and Porter supplied a 30-ton steam road-roller to the borough authorities of Liverpool, where it has been ever since at work to the satisfaction of the engineer. Subsequently to this, Messrs. Moreland and Son hired to the Government the steam roller now [1870] in use in Hyde Park. In 1867 steam road-rollers began to be introduced into Hanover. The manufacture of the rollers on the Ballaison construction, was taken up towards the end of 1868 by Messrs. Manning, Wardle, and Co., of Leeds; and, at the beginning of last year, a paper giving a description of the steam-roller used in Paris, was brought before the Institution of Mechanical Engineers, Birmingham, by M. Gellerat, of Paris, eliciting an interesting discussion on the whole subject.

APPENDIX III.

Extract from the Report of Colonel Haywood, Engineer and Surveyor to the Commissioners of Sewers.

CITY OF LONDON.

Condition of Wood and Asphalte Carriage-way Pavements in the City of London on 1st of February, 1877.

ASPHALTE.

Situation.	Pavement laid by	Date when laid	Area, Yards.	Time down.		Repairs done since it was laid new (from Inspector's Report).	Present condition.
				Yrs.	Mos.		
Aldgate	Limmer Asphalte Paving Co.	April, 1874	1,051	2	9	About one-third recently relaid	Indifferent condition; repairs constantly needed.
Bow Lane	Val de Travers Asphalte Paving Co.	Sept., 1873	425	3	4	None	Good.
Bread Street	Do.	June, 1875	362	1	7	None	Good.
Carter Lane	Barnett's Patent Asphalte Paving Co.	Sept., 1872	757	4	4	Several repairs by Barnett and Val de Travers Companies	Fair condition; but a few short holes.
Castle Street	Limmer Asphalte Paving Co.	Nov., 1873	567	3	2	Few repairs	Fair condition; channels slightly worn.

APPENDICES. 325

CHEAPSIDE	Val de Travers Asphalto Paving Co.	Sept. to Dec., 1870	6,933	6 1	Considerable repairs	Good; slight undulations of surface at places.
CORNHILL	Limmer Asphalte Paving Co.	Mar., 1872	3,333	4 10	Very much repaired	The surface for Asphalto is irregular, and signs of wear generally.
FENCHURCH STREET - Gracechurch Street to Railway Place	Barnett's Patent Asphalte Paving Co.	June, 1873	3,869	3 7	Most extensively repaired by Barnett and Valde Travers Companies	General surface fair, but many hollow places, the result of repairs.
FENCHURCH STREET - Railway Place to Aldgate	Société Française des Asphaltes	Sept., 1874	1,764	2 4	Extensively repaired	Good.
FETTER LANE (North end) Bartlett's Passage to Holborn	Brunswick Rock Asphalte Paving Co.	Oct., 1876	339	0 3	Few repairs	Channels slightly worn, otherwise good.
FETTER LANE (Middle) Bartlett's Passage to Fleur-de-Lis Court	Val de Travers Asphalte Paving Co.	Oct., 1876	1,607	0 3	None	Good.
FRIDAY STREET	Do.	Dec., 1876	125	0 1	None	Good.
FINSBURY PAVEMENT	Do.	Aug., 1871	3,861	5 5	Few repairs	Good.
FINCH LANE	Limmer Asphalto Pav. Co.	Sept. 1873	227	3 4	None	Good.

ASPHALTE—continued.

Situation.	Pavement laid by	Date when laid.	Area, Yards.	Time down. Yrs. Mos.	Repairs done since it was laid new (from Inspector's Report).	Present condition.
George Yard	Val de Travers Asphalte Paving Co. (Mastic Asphalte)	April, 1871	232	5 9	None	Good.
Gracechurch Street	Val de Travers Asphalte Paving Co.	July, 1871	2,659	5 6	Considerably repaired	Fair condition.
Great Bell Alley	Do.	May, 1876	82	0 8	None	Good.
Gresham Street	Do.	Mar., 1874	3,888	2 10	None	Good.
King's Arms Yard (Moorgate Street)	Do.	May, 1876	99	0 8	None	Good.
King Street (Cheapside)	Société Française des Asphaltes	Nov., 1874	1,299	2 2	Considerable repairs	Water lies close to kerb. Asphalte worn from Trump Street to Cheapside, surface otherwise fair.
King William Street Between Nicholas Lane and Clement's Lane	Val de Travers Asphalte Paving Co.	Sept., 1873	358	3 4	Few repairs	Good.

APPENDICES. 327

King William Street From thirty yards north of Nicholas Lane to Lombard Street	Do.	Nov., 1873	1,439	3 2	Few repairs	Good.
King William Street From Nicholas Lane thirty yards northward	Do.	Feb., 1874	299	2 11		
Lawrence Lane	Do.	May, 1874	482	2 8	None	Good.
Lombard Street	Limmer Asphalte Paving Co.	May, 1871	1,153	5 8	Considerable repairs	Surface for the most part regular.
London Wall	Val de Travers Asphalte Paving Co.	Oct., 1871	3,181	5 3	Very few repairs	Good.
Lothbury	Barnett's Patent Asphalto Paving Co.	Oct., 1872	2,384	4 3	Much repaired by Barnett and Val de Travers Companies	Generally good, but undulations between Coleman Street and Oboliak; requires frequent repairs.
Mansell Street	Val de Travers Asphalte Paving Co.	Oct., 1871	734	5 3	Few repairs	Good.
Mansion House Street St. Mildred's Court to Cornhill	Do.	Sept., 1870	3,043	6 4	Considerable repairs	Fair.
Milk Street	Do.	Mar., 1871	297	5 10	None	Good.

ASPHALTE—continued.

Situation.	Pavement laid by	Date when laid.	Area, Yards.	Time down. Yrs. Mos.	Repairs done since it was laid new (from Inspector's Report).	Present condition.
MINCING LANE	Limmer Asphalte Paving Co.	Aug., 1873	955	3 5	About twenty yards relaid during past year	Good.
MOORGATE STREET Lothbury to Great Bell Alley	Barnett's Patent Asphalte Paving Co.	Oct., 1871	1,057	5 3	Very much repaired by Barnett and Val de Travers Companies	Water lies along the east side of street; requires to be entirely relaid.
MOORGATE STREET From Great Bell Alley to near Coleman Street Buildings	Limmer Asphalte Paving Co.	Sept., 1871	1,038	5 4	Constantly being repaired	Very indifferent; requires repairs weekly.
MOORGATE STREET From Coleman Street Buildings to Ropemaker Street	Val de Travers Asphalte Paving Co.	Aug., 1871	1,027	5 5	Few repairs	Fair.
NEW BROAD STREET. Wormwood Street to Liverpool Street	Do.	Mar., 1871	879	5 10	Few repairs	Good.
NEWGATE STREET	Limmer Asphalte Paving Co.	July, 1876	4,161	0 6	About one-third relaid	Surface smooth, but is only kept in condition by constant repairs.

APPENDICES. 329

Street	Contractor	Date				Repairs	Condition
NEWGATE STREET (Western end)	Grindley and Co.	Do.	276	0	6	Nearly whole area relaid two or three times	Smooth, but repairs are constantly needed.
NICHOLAS LANE	Val de Travers Asphalte Paving Co.	June, 1875 Oct., 1876	137 292	1 0	7 3	None	Good.
OLD BAILEY (Middle)	Do.	May, 1871	402	5	8	Few repairs	Fair.
OLD BROAD STREET Threadneedle Street to Wormwood Street	Do.	May, 1874 Mar., 1871	217 2,792	2 5	8 10	Few repairs	Good.
OLD JEWRY	Limmer Asphalte Paving Co.	Oct., 1873	761	3	3	—	Good.
POULTRY	Val de Travers Asphalte Paving Co.	Sept., 1870	1,005	6	4	—	Fair condition. This pavement has been subject to much disturbance, owing to the improvements in the Street, and must be renewed this year, owing to alterations in the lines of footway and carriageway.
PRINCES STREET (Southern end)	Val de Travers Asphalte Paving Co.	Nov., 1872	711	4	2	Few repairs	Good.
PRINCES STREET (Middle)	Société Française des Asphaltes	July, 1872	327	4	6	Few repairs	Fair condition.

330   APPENDICES.

ASPHALTE—*continued.*

Situation.	Pavement laid by	Date when laid.	Area, Yards.	Time down.	Repairs done since it was laid new (from Inspector's Report).	Present condition.
				Yrs. Mos.		
Princes Street (Northern end)	Montrotier Asphalte and Cement Concrete Paving Co.	Aug., 1872	346	4 5	Considerable repairs. More than half is Val de Travers	Fair.
Queen Street Cheapside to Pancras Lane	Val de Travers Asphalte Paving Co.	April, 1871	312	5 9	Few repairs	Good.
Queen Street Pancras Lane to Queen Victoria Street	Do.	Sept., 1874	628	2 4	None	Good.
Rose Street	Val de Travers Asphalte Paving Co.	Oct., 1870	155	6 3	No repairs	Good.
Russia Row	Do.	Mar., 1871	80	5 10	None	Good.
St. Ann's Lane	Do.	Oct., 1873	452	3 3	Few repairs	Good.
Telegraph Street	Do.	May, 1876	133	0 8	None	Good.
Threadneedle Street Corner of Old Broad Street to West end of Merchant Tailors' Hall	Do.	May, 1869	485	7 8	Few repairs	Good.

APPENDICES. 331

Street	Contractor	Date			Repairs	Condition
THREADNEEDLE STREET Crown Court to Bishopsgate Street	Do.	Dec., 1871	439	5 1	Few repairs	Surface slightly irregular owing to traffic travelling on the same lines.
THREADNEEDLE STREET West end of Merchant Tailors' Hall to Crown Court	Do.	Jan., 1872	281	5 0	Few repairs	Do.
THREADNEEDLE STREET Princes Street to Old Broad Street	Do.	July, 1876	2,541	0 6	None	Do.
WOOD STREET Fore Street to Gresham Street	Do.	Sept., 1871	1,492	5 4	Few repairs	Good.
WOOD STREET Gresham Street to Cheapside	Do.	Sept., 1873	671	3 4	None	Good.

WOOD.

Street	Contractor	Date			Repairs	Condition
ALDERSGATE STREET	Improved Wood Pavement Co.	Sept., 1874	6,884	2 4	Much repaired	Surface level and good, but the joints open in many places.
ANGEL STREET	Do.	Sept., 1874	504	2 4	None	Good.
BARBICAN (Western end)	Do.	Nov., 1875	606	1 2	None	Good.
BARBICAN (Eastern end)	Asphaltic Wood Pavement Co.	Oct., 1875	785	1 3	Few repairs	Good, but surface slightly uneven.

WOOD—continued.

Situation.	Pavement laid by	Date when laid.	Area, Yards.	Time down. Yrs. Mos.	Repairs done since it was laid new (from Inspector's Report).	Present condition.
BARBICAN (Middle)	Carey	Oct., 1875	800	1 3	None	Good.
BARTHOLOMEW LANE (North end)	Do.	Jan., 1872	468	5 0	Few repairs	Surface generally level, but many joints open.
BARTHOLOMEW LANE (South end)	Improved Wood Pavement Co.	Dec., 1871	392	5 1	None	Level, but joints open.
BIRCHIN LANE (South end)	Carey	June, 1876	64	0 7	None	Good.
BISHOPSGATE ST. WITHOUT	Improved Wood Pavement Co.	Nov., 1874	7,955	2 2	Much repaired	Northern end bad, and must be relaid; some open joints, otherwise the condition is very fair.
BLOMFIELD STREET	Do.	Oct., 1875	1,394	1 3	None	Good.
BRIDEWELL PLACE	Do.	May, 1875	871	1 8	None	Good.
CANNON STREET St. Paul's Churchyard to South-west corner of Abchurch Lane	Carey	Sept., 1874	9,051	2 4	None	Good.

APPENDICES. 333

Cannon Street South-west corner of Abchurch Lane to North-west corner of Laurence Pountney Lane	Asphaltic Wood Pavement Co.	July, 1874	299	2 6	None	Good.
Cannon Street Laurence Pountney Lane to North-east corner of Nicholas Lane	Cartwright	June, 1874	306	2 7	Few repairs	Fair, but getting out of condition.
Cannon Street Nicholas Lane to King William Street	Mowlem & Co.	Sept., 1873	377	3 4	None	Good.
Carthusian Street	Do.	July, 1875	364	1 6	None	Good.
Chancery Lane	Improved Wood Pavement Co.	Nov., 1876	707	0 2	None	Good.
Charlotte Row	Do.	Sept., 1876	794	0 4	None	Good.
Coleman Street	Patent Ligno-Mineral Paving Co.	June, 1875	2,291	1 7	None	Good.
Duke Street	Mowlem & Co.	June, 1873	676	3 7	Few repairs	Fair; a short hole.
Eldon Street	Improved Wood Pavement Co.	Aug., 1875	1,471	1 5	None	Good.
Falcon Street	Mowlem & Co.	July, 1875	504	1 6	None	Good.
Falcon Square	Do.	July, 1875	589	1 6	None	Good.

WOOD—continued.

Situation.	Pavement laid by	Date when laid.	Area, Yards.	Time down. Yrs. Mos.	Repairs done since it was laid new (from Inspector's Report).	Present condition.
Fetter Lane (South end) Fleet Street to Fleur-de-Lis Court	Wilson	Oct., 1876	419	0 3	None	Good.
Fore Street	Patent Ligno-Mineral Paving Co.	Dec., 1874	3,822	2 1	Few repairs	Good.
Gracechurch Street	Do.	July, 1875	440	1 6	Relaid and part new	Fair, but wearing fast on the western side.
Great Tower Street and Seething Lane	Do.	Aug., 1873	448	3 5	Much repaired	Good.
Holborn	Do.	Jan., 1877	3,877	0 1	None	Good.
Houndsditch (Western half)	Carey	May, 1874	1,826	2 8	None	Fair.
Houndsditch (Eastern half)	Mowlem & Co.	May, 1874	1,832	2 8	None	Signs of wear on breast and channel.
Jewry Street	Carey	Jan., 1872	253	5 0	Few repairs	Blocks wearing round; surface generally indifferent.

APPENDICES. 335

...	Improved Wood Pavement Co.	Nov., 1876	1,126	0 2	None	Good.
KING EDWARD STREET	Do.	Nov., 1874	1,160	2 2	Few repairs	Good.
KING WILLIAM STREET	Do.	Aug., 1872	3,446	4 5	Greater portion re-laid with new blocks in 1876	Good.
KING WILLIAM STREET and ADELAIDE PLACE	Do.	Jan., 1873	2,620	4 0		
LEADENHALL STREET (Western half)	Improved Wood Pavement Co.	July, 1876	2,177	0 6	None	Good.
LEADENHALL STREET (Eastern half)	Henson's Street Paving Co.	Aug., 1876	1,788	0 5	None	Fair.
LITTLE BRITAIN	Improved Wood Pavement Co.	Oct., 1876	1,534	0 3	None	Good.
LONDON HOUSE YARD	Do.	April, 1874	43	2 9	None	Good.
LUDGATE HILL	Do.	Nov., 1873	2,824	3 2	Much repaired	Needs relay of the whole surface.
NEW BRIDGE STREET	Do.	Aug., 1876	3,817	0 5	None	Good.
OLD BAILEY (South end)	Do.	Jan., 1874	440	3 0	Few repairs	Fair.

WOOD—continued.

Situation.	Pavement laid by	Date when laid.	Area, Yards.	Time down. Yrs. Mos.	Repairs done since it was laid new (from Inspector's Report).	Present condition.
QUEEN STREET. Queen Victoria Street to Upper Thames Street	Asphaltic Wood Pavement Co.	July, 1876	2,169	0 6	None	Good.
ST. MARY AXE	Improved Wood Pavement Co.	July, 1875	1,662	1 6	None	Good.
ST. BRIDE STREET	Asphaltic Wood Pavement Co.	Nov., 1876	1,774	0 2	About 150 yards relaid	Good.
ST. MARTIN'S-LE-GRAND	Improved Wood Pavement Co.	Sept., 1874	2,650	2 4	None	Good, but joints open at one or two places.
ST. PAUL'S CHURCHYARD. North-west corner to South-west corner of Cathedral	Do.	Mar., 1874	2,182	2 10	None	Good.
ST. PAUL'S CHURCHYARD. South-west corner of Cathedral to Cheapside.	Gabriel	Jan., 1876	4,907	1 0	None	Good.

St. Swithin's Lane	Improved Wood Pavement Co.	June, 1876	452	0 7	None	Good.
Silver Street	Mowlem & Co.	July, 1875	571	1 6	Few repairs	Surface good, but water lies slightly next kerb on South side.
South Place	Improved Wood Pavement Co.	July, 1876	1,753	0 6	None	Good.
Walbrook	Mowlem & Co.	April, 1875	455	1 9	None	Good.
Watling Street	Improved Wood Pavement Co.	Sept., 1876	2,100	0 4	None	Good.
Wormwood Street	Mowlem & Co.	June, 1875	689	1 7	None	Good.

Note.—Fleet Street was repaved with Wood Pavement in August, 1877:—A portion, from Ludgate Circus to Bouverie Street, consisting of Asphaltic Wood Pavement; and the remainder, towards Temple Bar, of Henson's Wood Pavement.—D. K. C.

INDEX.

ASPHALTE for pavement, 17, 130; artificial asphalte, 131; wheel-tracks, 134, 214, 242
Asphalte, Barnett's liquid iron, pavement, 245, 249, 254, 256, 258, 259; cost, 259; wear, 254, 256
Asphalte, Bennett's foothold metallic, pavement, 246
Asphalte, Lillie's composite, pavement, 247
Asphalte, Limmer mastic, pavement, laid in the City of London, 244, 249, 254, 256, 258, 259; cost, 237, 256, 259; wear, 254, 256
Asphalte, Maestu compound, pavement, 246
Asphalte, Montrotier compressed, pavement, 246, 249, 254, 256; cost, 246; wear, 254, 256
Asphalte, patent British, pavement, 245
Asphaltes, Société Francaise des, pavement, by, 246, 249, 254
Asphalte, Stone's slipless, pavement, 246
Asphalte, Trinidad, pavement, 245
Asphalte, Val de Travers, pavement, laid in Paris, 242; in the City of London, 243, 249, 252, 256, 258, 259; cost, 237, 242, 244, 259; wear, 250, 252, 256; average wear, 257
Asphalte joints, granite pavement with, in Manchester, 199; in the City of London, 247

Asphalte pavement, wear of, 250, 252, 254, 256
Asphalte pavements, 242:—Val de Travers asphalte laid in Paris, 242; asphalte pavements in the City of London, 243; cost, 259; in Manchester, 259
Asphalte-granite pavement, in Manchester, 199, in the City of London, 247
Asphaltic wood pavement, 229, 235, 236, 237, 238

BALFOUR, MR. "D., on the cost of maintenance of macadam roads at Sunderland, 161
Barnett's liquid iron asphalte pavement, 245. (See *Asphalte, Barnett's*, &c.)
Bennett's foothold metallic asphalte pavement, 246
Birmingham:—Wear and cost of macadamised roads, by Mr. J. P. Smith, 155; comparative cost of macadam and paved roads, by Mr. W. Taylor, 158; cleansing the streets of, 273
Blackfriars Bridge, stone pavement, 173, 187
Bode, Baron de, on wood pavement in Russia, 13
Bokeberg, Herr E., on the interspaces in broken metal, 145
Boning rods, 134
Boulder pavement, 11; in the City of London, 173; in Liverpool, 195; in Manchester, 198

British, patent, asphalte pavement, 245
Browne, Major James, on mountain roads, 283
Burgoyne, Sir John, on road-rolling, 11; appendix, 301

CAREY'S wood pavement, 17, 217, 234, 236, 237, 238, 239
Cast-iron paving, 261
Cellular iron pavement, 262
Chalk as a building material for roads, 101
Charié-Marsaines on performance of Flemish horses, 299
Clark, D. K., on the performance of horses, 300
Cleansing of roads and streets, 268. (See *Roads and Streets, Cleansing of*)
Commercial Road, stone tramways in, 208
Concrete as a foundation for roads, by Mr. Hughes, 92; and by Mr. Penfold, 92
Concrete pavement, McDonnell's adamantean, 247; Mitchell's, 263
Concrete roads, by Mr. Joseph Mitchell, 162
Construction of roads, 40, 90; earthwork, 40; working plan, 40; slopes, 43, 46; drainage, 46, 70; embankments, 47; catch-water drains and cross drains, 48; roads on the side of a hill, 48; side-cuttings, 49; spoil-bank, 50; maximum gradients, 63, 65; minimum gradients, 64; foundation and superstructure, 90; three kinds of foundation, 91; concrete foundation, by Mr. Hughes, 92; and by Mr. Penfold, 92; foundation of pavement, by Mr. Telford, 94; broken stone first recommended by Macadam, 97; Mr. Hughes on broken stone covering, 99; chalk as a binding material, 101; Mr. Walker on iron scraps, as a binding material, 102; foundation for paved streets, 105; materials employed in the construction of roads and streets, 123; modern macadamised roads, 134
Contour lines and maps, 24
Cost of asphalte pavements, 259, 264, 265
Cost of macadamised roads, 151, 156, 158, 161, 162, 163
Cost of stone pavements:—First cost and repair of streets in the City of London, 171, 183, 184, 188; Euston pavement, 176; experimental paving in Moorgate Street, 177; London Bridge, 186; Blackfriars Bridge, 187; Liverpool, 194, 195; comparative costs, 265
Cost of Yorkshire paving, 265
Cost of wood pavements, 217, 236
Crossings, 108
Cunningham, Mr. J. H., on the cost of construction of macadam roads near Edinburgh, Glasgow, and Carlisle, 162
Curbs, 106

DERBY, cost of macadamised streets in, by Mr. E. B. Ellice-Clark, 158, 161
Debauve, resistance to traction on common roads, 295
Dobson, Mr., on mountain roads, 284
Dockray, Mr., his opinion of the Euston pavement, 176
Drainage, 46, 48, 74
Dumas on French roads, 19; on gradients of roads, 63; on modern macadamised roads in France, 165
Dupuit, M., on the gradients of roads, 64; on the width of surface of tyres in contact, 144; experiments on resistance to traction, 295

EARTHWORK, 40
Eastons and Anderson on resistance of carts and waggons, 297
Edgeworth on the construction of roads, 9

Edinburgh, Glasgow, and Carlisle, cost of construction of macadam roads near, by Mr. J. H. Cunningham, 162
Ellice-Clark, Mr. E. B., on the cost of macadamised streets in Derby, 158, 161; estimate cost for paving and maintaining streets in Derby, 160
Embankments, 47
Estimates, taking out quantities for, 114
Exploration for roads, 21; laying out, 21, 36: contour-lines and maps, 24; working sections, 37

FAREY, MR. JOHN, on the work of horses and the wear of roads, 140
Fences, 110. (See *Hedges and Fences*)
Footpaths, 77, 108; stones used for, 128
Foster, P. Le Neve, Jun., on tramways in Milan, 208
Foundations for pavements, 104, 105
Foundation of roads, 90. (See *Construction of Roads*)
France, roads in, 18, 165. (See *Roads in France*)

GABRIEL'S wood pavement, 233, 235
Gradients, maximum, 63, .65; minimum, 64
Granite, 122; comparative wear, 126
Granite pavement, artificial, Poletti and Dimpfl's, 262
Granite tramways, 208. (See *Stone Tramways*)
Greywacké, 127

HARRISON'S wood pavement, 231
Haywood, Colonel, Reports of, to the Commissioners of Sewers of the City of London, 170; earliest carriage-way pavements in the City of London, 171; pavements in 1848, length of carriage-way in 1851 and 1866, three-inch granite sets made the best pavement, 174; experimental paving in Moorgate Street, 177; traffic in the City of London in 1850, 1857, 1865, and 1871, 181; duration of three-inch set pavements, 182; cost, 184; on the wear and cost of paving for London Bridge, 185; estimate of durability and cost of pavements in principal streets of the City, 187; typical sections of a fifty-feet street, 188; table, showing the condition of wood and asphalte carriage-way pavements in the City of London, on 1st February, 1877, 324
Hedges and fences, 110; stone fences, 110; post and rail fence, 111; quickset hedge, 111; Professor Mahon on fences, 112; Sir John Macneil on the evil of close fences, 112; Mr. Walker on the same, 113
Henson's wood pavement, 231, 237
Hope, D. T., his experiments on wear of wood in pavements, 132
Horse tracks, 1
Hughes, Mr., on the form of the bed of a road, 73; on broken stone covering, 99; concrete foundation by, 92

IMPROVED wood pavement, 223, 234, 236, 238
India, road-making in, 282; the Grand Trunk Road, 285
India, stones in, 127
Iron scraps as a binding material for roads, 102

KELSEY, MR., on cost for a pair of early stone pavements in the City of London, 1
Kunker, 128

LATERITE, 127
Lee, Mr., experiments in cleaning the streets of Sheffield,

Ligno-mineral pavement, 228, 234, 237, 238
Lillie's composite pavement, 247
Limmer mastic asphalte pavement, 244. (See *Asphalte, Limmer*, &c.)
Lisle, Count de, his wood pavement, 16
Liverpool, stone pavements of, 193. (See *Stone Pavements*)
London, boulder pavement, 11; Mr. Telford's system of pavement, 12; Macadam's pavement, 13; construction, wear, and cost of metropolitan roads, 134, 138, 151; cleansing and watering the streets of, 277, 278
London, City of, first Act for paving, 11; paved with boulders, 11; macadamised streets, 13; stone pavements, 169; wear of granite pavements, 202; wood pavements, 217; asphalte pavements, 243. (See *Stone Pavements, Wood Pavements, Asphalte Pavements*)
London Bridge, construction, and wear of stone pavement, 185
Lovick, Mr. T., his experiments in cleansing streets, 278

MACADAM in Manchester, 201
Macadam, James L., on the old roads, 7; on the principle of a good road, 7, 8; on the section of roads, 69; use of broken stone, 97; on the width of tyres, 144; on the annual wear of metalled roads, 147
Macadamised roads, leading principle of, 7; total length of, in 1868-69, 10; rolling, 10; adopted in France, 19; not suited for heavy traffic, 104; stones suited for, 127; modern macadamised roads, 134. (See *Construction of Roads*)
Macadamised roads, modern, construction of, 134; first-class metropolitan roads, 134; second-class metropolitan roads, 135; country roads, 135; construction proposed by a Committee of the Society of Arts, 137; wear, 138; disadvantage of elasticity, 138; relative wear due to the action of horseshoes and of wheels, 140, 144; rounded tyres injurious, 141; width of surface in contact with tyres, 144; interspaces in broken stone, 145; analysis of crust of macadamised road, by Mr. Mitchell, 146; annual wear, 147; rules for wear, 148; maintenance of roads in the metropolis, 151; suburban highways, 153; local roads, 155; Birmingham, 155; Derby, 158; Sunderland, 161; districts near Edinburgh, Glasgow, and Carlisle, 162; macadamised roads in France, 165; in Manchester, 201. (See *Wear and Cost*)
Macneil, Sir John, experiments by, on resistance to traction, 52, 298; on inclination of roads, 63; on close fences, 112; on elasticity of the road, 138; on the weight of vehicles and width of tyres, 142; experiments on resistance of granite tramways, 208
Macstu compound asphalte, 246
Mahan, Professor, on side slopes, 43; on the angle of repose, 61; on maximum gradients, 63; on the form of cross section, 72; on the height of fences, 112
Manchester, asphalte pavements in, 259
Manchester, cleansing the streets of, 270, 276
Marshy soils, brushwood substructure in, 77
Materials employed in the construction of roads and streets, 122:—stones, 122; granite, 122; its crushing resistance, 124; absorbent power of stones, 125; trap rocks, 125; comparative wear of stones, 126; Mr. Walker on wear of granites, 126; greywacké, 127; stones suited for macadam, 127; stones in India,

127; stones for footpaths, 128; for curbs, 130; asphalte, 130; artificial asphalte, 131; wood, 131; its crushing resistance, 132; Mr. D. T. Hope's experiments on wear, 132
McDonnell's adamantean concrete pavement, 247
Metalled roads. (See *Macadamised Roads*)
Metallic paving, 261, 262
Metropolitan compound metallic paving, 261
Metropolitan Wood Pavement Company, their pavement, 16
Milan, stone tramways in, 208
Mitchell, Mr. Joseph, on the interspaces in broken stone, 145; analysis of the cost of a macadam road, 146; concrete road, 162; concrete pavement, 263
Montrotier compressed asphalte pavement, 246. (See *Asphalte, Montrotier*, &c.)
Morin's experiments on resistance to traction, 51, 294
Mountain roads, 283
Mowlem's wood pavement, 232, 235, 236, 237
Mud, removal of, 82; composition of, 268, 275

NEWLANDS, MR., on the streets of Liverpool, 193
Norton's wood pavement, 232, 235

PACK-HORSES, 1

Paget, F. A., his data, 149, 151, 309
Paris, cleansing the streets of, 281.
Pavements, carriage-way, comparison of:—cost, 264; slipperiness, 259, 265; convenience, 267; cleanliness, 281
Pavement, boulder. (See *Boulder Pavement*)
Pavement, cast-iron, 261
Pavement, cellular iron, 262
Pavement, compound wood and stone, 262
Pavement, granite, estimate cost of, in Birmingham, 158; in Derby, 160, 161
Pavement, granite, artificial, 262
Pavement, metallic, 261, 262
Pavement, stone. (See *Stone Pavement*)
Pavement, wood. (See *Wood Pavement*.)
Paved roads and streets, 104; foundations, 104; mode of preparing surface for pavement, 105; construction of foundation, 105; stone sets, 105; curb, 106; pavement for inclined streets, 108; side walls and crossing places, 108
Penfold, Mr., on repairs of roads, 81; concrete foundation by, 92
Pinchbeck, Mr. George, on cost of maintaining suburban highways, 153
Poletti and Dimpfl's artificial granite pavement, 262
Polonceau on road rolling, 20
Provis, Mr., on elasticity of roads, 138

QUANTITIES for estimates, taking out, 114

READING, cleansing the streets of, 277
Redman, Mr., on the wear of macadam in Commercial Road, 148
Repairing and improving roads, 79; improvement of the surface, 79; lifting the road, 80; Mr. Penfold on repairs, 81; removal of mud, 82; tools or implements formerly employed, 83; scraping machines, 89; Whitworth's sweeping machine, 89
Resistance to traction on common roads, 51, 290; M. Morin's experiments, 51; Sir John Macneil's experiments, 52; rules for resistance, 55; influence of inclines, 56; angle of repose, 61; Sir John Macneil on gradients, Professor Mahan on gradients, M. Dumas on gradients, 63;

INDEX. 343

M. Dupuit on gradients, 64;
minimum longitudinal slope, 64.
Rolling resistance, 290; conclusions and data of M. Dupuit,
295; M. Debauve, 295; M.
Tresca, 296; Messrs. Eastons
and Anderson, 297; Sir John
Macneil, 298; formula, 299;
M. Charié - Marsaines' data,
299; Mr. D. K. Clark's data,
300
Resistance of granite tramways,
208
Road, modern country, 136
Road-rolling, 10; in France, 20,
167
Roads and streets, cleansing of,
268; composition of mud and
detritus, 268, 275; cleansing
streets of Manchester, 270, 276;
Whitworth's machine, 271;
cleansing streets of Salford,
273; of Birmingham, 273; of
Reading, 277; watering the
streets of London, 277; Mr. T.
Lovick's experiments, 278; Mr.
Lee's experiments at Sheffield,
279; cleansing in Paris, 281;
M. Tailfer's machine, 281
Roads, construction of, 40, 90.
(See *Construction of Roads*)
Roads, exploration for, 21. (See
Exploration for Roads)
Roads in France:—old roads, 18;
Trésaguet's roads, 19; Macadam's system adopted, 19;
horse-roller adopted, 20; modern macadamised roads in
France, 165; M. Dumas on
their construction, 166; rolling
and watering, 167
Roads, metalled or macadamised.
(See *Macadamised Roads*)
Roads, mountain, 283
Roads, old country, 3; Mr. Macadam's opinion of, 7
Roads, paved, 104. (See *Paved
Roads and Streets*)
Roads, repairing and improving,
79. (See *Repairing and Improving Roads*)
Roads, section of, 65. (See *Section of Roads*)

SALFORD, cleansing the streets
of, 273
Sandstone, strength and absorbent
power of, 125; laterite in India,
127; for footpaths, 128; composition, weight and strength,
129; comparative durability,
129
Scraping machines for macadam
roads, 89
Section of roads, 65; limits of
gradients, 66; width and transverse section of roads, 68; Mr.
Macadam on the section of
roads, 69; Mr. Walker on the
section of roads, 70; and on
drainage, 70; best form of section, 71; section of the bed,
72; Mr. Hughes on the form
of the bed, 73; drainage of the
road, 74; footpaths, 77; brushwood substructure in marshy
soils, 77, 102; imperfect sections, 80
Sections, working, for roads, 37
Scyssel asphalte, 133, 131
Sheffield, cleansing the streets at,
279
Side-cuttings, 49
Side walks, 108
Slopes, 43, 46
Smith, Mr. J. P., on macadamised
roads in Birmingham, 155
Société Française des Asphaltes,
246, 249, 254
Society of Arts, Report of Committee of, on Traction on Roads,
208
Spoil-bank, 50
Stead's wood pavement, 15
Stone pavement, 169; City of
London, 169; early paving,
169; pavement of King William Street, 170; table of early
pavements, 171; table of cost
for repairs of early pavement,
172; introduction of three-inch
sets; extent of pavement in
the City, 173, 174; the Euston
pavement, 175; experimental
paving in Moorgate Street, 177;
granites that have been tried,
179; rotation of paving stones

180; traffic in the City, 181; duration of pavements, 183, 187; cost, 184, 187; London Bridge, 186; Blackfriars Bridge, 187; typical sections and plans of a fifty-feet street, 188 : Southwark Street, 191. Liverpool, 193; extent of pavement in 1851, 193 ; cost for construction of set pavements, 194; of boulder pavements, 195; and of macadam, 196; cost of maintenance of pavements, 197. Manchester, 198 ; boulder pavement, 198 ; construction of set pavement, with asphalte jointing, 199; cost of pavements, 200 ; wear of granite pavements in the City of London, 202

Stone tramways, 208 :—Commercial Road, by Mr. Walker, 208 ; resistance on, 208 ; in Northern Italy, described by Mr. P. Le Neve Foster, Jun., 208 ; prices in Milan, 212 ; wheel-tracks of asphalte, 214

Stones used in the construction of roads, 122. (See *Materials employed*)

Stone's slipless asphalte pavement, 246

Stone's wood pavement, 233

Streets, typical :—fifty-feet street for the City of London, 188 ; Southwark Street, 191

Sunderland, cost of maintenance of macadam roads at, by Mr. D. Balfour, 161

Sweeping machines, Whitworth's, 89, 271; Tailfer's, 281

TAILFER'S sweeping machine, 281

Taylor, Mr. W., on the Euston pavement, 175; comparative cost of macadam and paved roads in Birmingham, 158

Telford, his early experience, 9 ; his system of road-making, 10, 94 ; his system of pavement for London, 12; on the wear of roads, 141

Tools or implements formerly employed in the repair of roads, 83

Traction, resistance to, 51, 290. (See *Resistance to Traction*)

Tramway, stone, 208. (See *Stone Tramways*)

Trap rocks, 125

Trésaguet's roads in France, 18, 19

Tresca, M., on the resistance of an omnibus, 296

Trinidad asphalte pavement, 245

VAL DE TRAVERS asphalte, 130, 131

Val de Travers asphalte pavement, 242. (See *Asphalte, Val de Travers, Pavement*)

Vehicles on common roads in 1809, 7

WALKER, MR., on the section of roads, 70; on drainage, 70; on iron scraps as a binding material, 102; on close fences, 113; on the wear of granites, 126; introduction of three-inch sets by, 173; stone tramways in Commercial Road, 208

Watering streets. (See *Roads and Streets, Cleansing of*)

Wear of macadamised roads, 138, 140, 144, 147, 148, 155, 156, 158

Wear of pavements :—stone, 202; wood, 238 ; asphalte, 250, 252, 254, 256 ; holes and short holes in asphalte, 250, 253

Wear of stones, comparative, 126

Whitworth's sweeping machine, 89, 271

Wilson's wood pavement, 233

Wood, 131 ; its crushing resistance, 132

Wood and stone pavements, compound, 262

Wood pavements in the City of London, 234

Wood pavement in Russia, 13 ; in the United States, 14 ; Stead's pavement, 15 ; De Lisle's pavement, 16; general conditions of wood pavement, 215; Carey's wood pavement, 17, 217, 234, 236, 237, 238, 239; improved wood pavement, 223, 234, 237, 238; Ligno-mineral pavement, 228, 234, 237, 238; asphaltic wood pavement, 229, 235, 236, 237, 238; Harrison's wood pavement, 231; Henson's wood pavement, 231, 237; Norton's wood pavement, 232, 235 Mowlem's wood pavement, 232, 235, 236, 237; Stone's wood pavement, 233; Gabriel's wood pavement, 233, 235; Wilson's wood pavement, 233

YORKSHIRE paving, 191, 265

Young, Arthur, on the old roads, 3

THE END.

www.ingramcontent.com/pod-product-compliance
Lightning Source LLC
Chambersburg PA
CBHW030252240426
43673CB00040B/947